T0181923

Computer science to the Point

Boris Tolg

Computer science to the Point

Computer Science for Life Sciences Students and Other Non-Computer Scientists

 Springer

Boris Tolg
Department Medizintechnik
HAW Hamburg Fakultät Life Sciences
Hamburg, Germany

ISBN 978-3-658-38442-5   ISBN 978-3-658-38443-2   (eBook)
https://doi.org/10.1007/978-3-658-38443-2

This Springer imprint is published by the registered company Springer Fachmedien Wiesbaden GmbH, part of Springer Nature.
The registered company address is: Abraham-Lincoln-Str. 46, 65189 Wiesbaden, Germany

For my wife Christine and my sons Jean Luc and Mateo Nicolao.

Preface

The idea for the Life Sciences book series came about in December 2015, when I was in the process of developing new teaching concepts for my computer science lecture. While searching for suitable literature, I noticed that although there is a wide range of books, workshops, and online tutorials for the C++ programming language on the market, none of them meet the needs of my students.

Each of the offerings in itself, of course, serves its purpose and is sometimes more and sometimes less successful, but all of them assume that readers have an intrinsic motivation to become involved in computer science.

However, this motivation cannot necessarily be assumed for students who take computer science as a minor in an interdisciplinary degree program. As diverse as the composition of subjects in an interdisciplinary degree program is, so diverse are the careers of students who have completed these programs. Not all of them focus on software development, and some will probably never write another program after graduation.

In these courses, it is important to provide students with a different motivation as to why they should pursue computer science.

However, this problem does not only affect computer science, but all basic subjects of an interdisciplinary course of studies. Consequently, it makes sense to build up a common teaching concept that extends across different basic subjects. The idea is to define tasks that originate from the respective courses of study and then to break them down into problems of the basic subjects. In this way, the individual books can function on their own as textbooks for their field of application, but in interaction with the other volumes they can offer a self-contained description of solutions for subject-specific problems. You, the reader of these lines, thus have the possibility to concentrate only on the works that are really interesting for you. The subject-specific problems provide motivation to deal with the respective basic subject.

I am very happy that I was able to win over a number of colleagues to help me realize my idea. Thus, in addition to the book on computer science, other books are appearing with the subjects of physics and mathematics. The future will show whether the list will become even longer.

Of course, I would be very pleased if the teaching concept of the books would simplify your entry into the basic subjects. Therefore, I wish you much fun and success in reading the book and learning the basics of computer science.

I have tried to adapt the book to the best of my knowledge and belief to the needs of students in the minor subject. I would like to express my sincere thanks to two of my students, Ms. Lea Jungnickel and Ms. Sandra Kerstin Spangenberg, who read the book from a student's perspective and pointed out the incomprehensible passages to me.

Prof. Dr. Holger Kohlhoff and Prof. Dr. Jürgen Lorenz supported me in terms of content during the troubleshooting. I would also like to thank them from the bottom of my heart.

Reinbek   Boris Tolg
August 2018

Contents

Part I

Introduction

How to Work with This Book

1

This question certainly cannot be answered the same way for everyone, because each person reading this book has different prior knowledge and needs in terms of how knowledge should be imparted. Therefore, perhaps the first important thing is who should read this book:

The book is intended for people who are confronted with the subject of computer science and programming either in the course of studies, training, purely out of interest, or for other reasons. Presumably, computer science is a minor subject and is not the central topic of your interest. I assume that you have no or very little basic knowledge. I also assume that the subject of computer science has not played a significant role in your life to date. You may even have always been somewhat suspicious of the subject and the people who deal with it.

If you have the opportunity, try not to go through the book alone, but find a study group. First of all, it is more pleasant when several people support each other in working through a new and perhaps even unpopular topic. However, experience has shown that it also improves learning when they try to explain to each other what they have already understood. Whenever you are struggling for words when explaining something, you haven't really understood it yet and should look at the topic in question again.

The important thing is that you try to explain everything in your own words. If you do not have a study group, pets or an empty chair are also suitable.

How do you work with the book now?

The book is divided into three major sections. In the first part of the book, some basics are taught. You should read the chapter on *syntax diagrams* right at the beginning, because you will need this knowledge in the second part of the book.

The *Unified Modelling Language* (Object Management Group 2018), called UML for short, is a graphical language that will enable you to represent complex processes in

B. Tolg, *Computer science to the Point*, https://doi.org/10.1007/978-3-658-38443-2_1

programs simply and clearly. It consists of many different types of diagrams, but only three of them will be presented in this book. The use case diagrams document one of the first steps on the way to a new software project; they are easy to understand but are not needed until the third part of the book. The activity diagrams provide a detailed description of program flows and are used as early as the second part of the book. The class diagrams require knowledge of classes, so you should wait until you have completed the second part of the book to use them.

At the end of the first part you will find a short introduction to competence orientation and Bloom's taxonomy levels. There, it is explained how the exercises in this book are conceptualized and which conclusions you can draw for yourself.

The second part of the book explains the basics of the programming language *C++*. If you have no experience with *C++* you should work through this part of the book from front to back. The later chapters always build on the knowledge of the previous ones. Occasionally, however, there is content in a topic that you do not necessarily have to understand the first time you work through it. These chapters deepen the knowledge and are marked as advanced in the headings.

But before you start reading the second part of the book, you should get a development environment for the language *C++*. There are free development environments for every major operating system. The examples in this book were tested with the free *Microsoft Visual Studio 2017 Community* Development Environment (Microsoft 2017), which can be used on Windows systems. Other free development environments include *Code::Blocks* (Code::Blocks 2017) for Windows, *Xcode* (Apple Distribution International 2017) for *macOS,* and the platform-independent development environment *Eclipse CDT* (Eclipse Foundation 2017). However, this list is by no means exhaustive.

In the third part of the book, the knowledge from the previous two parts is used to face subject-related problems. There, it is explained what a more complex task might look like and how you can get to a practical implementation step by step.

Documentation of Languages and Programs

2

2.1   Syntax Diagrams

In computer science, syntax diagrams are used to graphically represent the structure of statements. The term syntax already suggests that it is about how words become sentences. However, syntax diagrams go one step further and are also used to first define what is meant by the term "word".

In Fig. 2.1, some terms are first defined that can be reused in other diagrams. The term to be defined is always placed above the diagram, followed by a colon.

The arrows in the diagram indicate the direction in which the diagram can be traversed. It is not allowed to move against the direction of the arrows. The first diagram on the left defines the term *lowercase*. It shows various paths leading from the left to the right edge of the diagram, each pointing to a circle. Within the circles, or more precisely within rectangles with rounded corners, characters or text are specified to be rendered exactly as defined in the diagram. In this case, they are all letters of the alphabet in their lowercase form.

The other two diagrams define analogously the meaning of the word *uppercase* as a letter of the alphabet in the capitalized form, or the word *digit as a* digit between 0 and 9.

After these terms have been defined, they can be used in further diagrams. In Fig. 2.2, the two terms *lowercase* and *uppercase* are used to define the term *letter.*

If terms are to be used that have already been defined in other diagrams, these terms are written in a simple rectangular box. Thus, according to this definition, the term *letter* denotes either an *uppercase letter* or a *lowercase letter*.

However, the syntax diagrams also allow more complex syntactic constructs to be described. The programming language *C++* allows you to specify names for functions and variables that can be used in programs. The names may consist of letters, digits and the

B. Tolg, *Computer science to the Point*,
https://doi.org/10.1007/978-3-658-38443-2_2

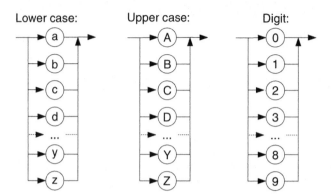

Fig. 2.1 Syntax diagram for the definition of lower and upper case letters as well as digits

Fig. 2.2 Syntax diagram for
the definition of a letter as
either a lowercase or uppercase letter

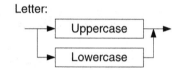

underscore. However, variable or function names must never begin with a digit. This structure is shown in Fig. 2.3 with the aid of a syntax diagram.

Starting from the left margin, in the first part of the diagram there is a choice between a *letter* and the underscore, which corresponds to the first allowed character of a variable name. In the second part, there is a choice between all three options. Just before the right edge of the diagram, there is an additional arrow that allows to return before the selection of the three options. This arrow represents an optional loop, so the selection can be repeated as many times as desired. In other words, the variable name can be of any length and it is possible to use a digit, a letter or an underscore for every character except the first.

For example, a valid variable name according to this definition would be *_TesT5,* while the variable name *1_test* would be invalid because it starts with a digit.

In this book, syntax diagrams are used at various points to illustrate the structure of *statements.* However, the term *statement* itself is not defined, although it is used in some diagrams. This is due to the fact that many language elements introduced in the second part of the book are *statements* that can be inserted at the appropriate place. In addition, many other language elements exist beyond the scope of the book that also fall under this term. Consequently, a complete definition of the term by a syntax diagram would be confusing and would not help to understand the language *C++.*

For the same reason, no syntax diagrams were prepared for some statements or data structures, such as the classes in Chap. 10. In these cases, however, the structure is clarified with the aid of examples.

Some syntax diagrams use the term block, which is defined in Fig. 2.4.

Variable name:

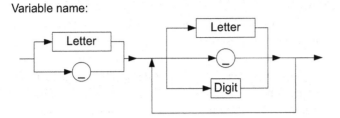

Fig. 2.3 Syntax diagram for the definition of function and variable names in

Fig. 2.4 Syntax diagram for the definition of a statement block

Block:

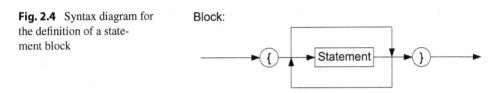

It is a sequence of statements enclosed by curly braces. However, within the curly braces there is also a path that leads past the statement. This means that the statement is optional and can therefore be omitted. In this case, the block consists of only one pair of curly braces.

2.2 Unified Modelling Language (UML)

The *Unified Modelling Language, or* UML for short, is a graphical language that can be used to model and describe software systems. Since complex systems are described through this modeling, the design and creation of the description of a software system is also called software architecture. The UML defines various types of diagrams and expressions that enable even non-computer scientists to understand the relationships in complex systems. The first version of UML was essentially driven by three people. Grady Booch, Ivar Jacobson and James Rumbaugh ("The Three Amigos") had initially developed their own modeling languages, which they later merged into the common language UML.

The specification and further development of UML is carried out by the *Object Management Group* (Object Management Group 2018).

In UML, a rough distinction is made between two types of diagrams, the structure diagrams, which take a static view of the individual components of a software and relate them to each other, and the behavior diagrams, which describe communication and dynamic processes. In the following, an overview of all diagrams of the UML is given with a short description. However, some of the diagrams describe relationships that require more experience than can be provided by reading this book. Therefore, the descriptions are only intended to provide a rough estimate.

- Structure charts
  - Class diagram – Models the relationships of classes and their interfaces.
  - Component diagram – Describes the relationships of more complex components (which can be described by class diagrams) and their interfaces.
  - Object diagram – Describes, among other things, the assignment of the attributes of a class to certain objects.
  - Profile Diagram – Allows you to define extensions for applying UML to specific systems, called profiles.
  - Composite Structure Diagram – Describes the internal parts of a complex component and their relationships to each other and to the outside world.
  - Deployment Diagram – Describes the distribution of software across multiple computers for complex systems.
  - Package Diagram – A flexible diagram that allows you to combine other diagrams or descriptions into one Package and describe connections to other Packages.
- Behaviour charts
  - Use Case Diagram – Gives an overview of the actors and with what goal they use a software.
  - Activity Diagram – Describes the individual actions and their relationships that must be performed when implementing a use case.
  - State Machine diagram – Describes the various states of a finite state machine.
  - Interaction diagrams
  - Sequence diagram – Describes which objects communicate with which messages. The focus is on the clear representation of the chronological sequence of the messages.
  - Communication Diagram – Also represents communication between objects, but with an emphasis on the relationships between objects.
  - Interaction Overview Diagram – Can combine elements from the activity diagram with those from other interaction diagrams to describe complex processes inside and outside components.
  - Timing Diagram – Represents the changes of states of objects on a timeline.

In larger software projects, the use of these diagrams is useful to better plan and document the software. The communication between different developer groups is also improved by the easily understandable diagrams. For the introduction to computer science, however, only three types of diagrams will initially be described and applied in more detail.

2.2.1   The Use Case Diagram

Use case diagrams are created at a very early stage of software development. They are used to approach the tasks of a software system by documenting what types of people will interact with the system. For each of these *actors, use cases* are then identified in which

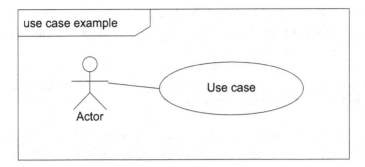

Fig. 2.5 Use case diagram with one actor and one use case

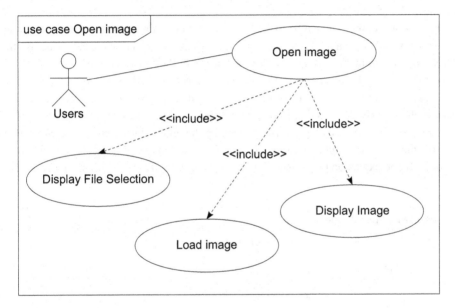

Fig. 2.6 Use case diagram for loading an image with partial use cases

they will interact with the software. In Fig. 2.5, an actor, represented by a stick figure, and a use case are documented.

The use case is always symbolized by an ellipse. The relationship between actor and use case is represented by a solid line called association.

All diagrams of the UML are enclosed by a solid frame, in the upper left corner of which an abbreviation for the diagram, as well as a description for the displayed content can be found. This description is always enclosed by a near-rectangular box whose lower right corner is indented. In this case, *use case* stands for the use case diagram and *example* for the short description. Instead of *use case,* the UML also allows the abbreviation *uc*.

In Fig. 2.6, a concrete use case for an image processing software is now to be described, which will additionally be extended by several partial use cases.

The use case described is intended to describe the opening of an image, so the name was set accordingly in the upper left corner of the diagram. The user (or users) of the software to be created was associated with this use case.

However, the *Open Image* use case is composed of several sub-aspects that may also be relevant in other use cases. Therefore, it may be useful to document these sub-aspects as individual sub-use cases. In this way, common aspects of different use cases can be highlighted.

In this example, the *Open Image* use case additionally includes the *Display File Selection, Load Image,* and *Display Image* use cases. The connection is represented in the diagram by a directed connection in the form of a dashed line with a black arrowhead. The arrowhead points from the parent use case to the sub use case. The connection must be marked with the note *<<include> >* .

Optional use cases can also be modeled using use case diagrams. For example, it may be necessary to display a help function when an appropriate selection is made. But also the selection of a certain file format can be an extension of the described use case.

In this case, the UML provides that the use case is traversed by a horizontal line. The upper part still contains the name of the use case, while so-called extension points can be listed in the lower part. This list is always introduced with the term *extension points.*

In Fig. 2.7 the use case diagram is extended by the optional use case *Select file format.*

First, the use case *Display file selection* was divided into two areas as described before. In addition, the extension point *Selection* has been added. This name is assigned so that it

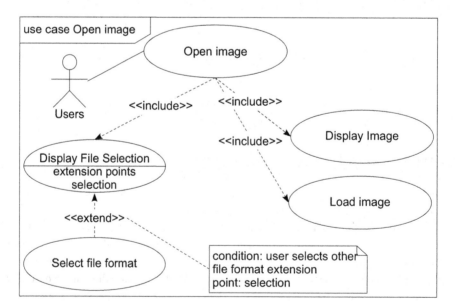

Fig. 2.7 Use case diagram for loading an image with partial use cases and extension for file format selection

can be referred to later at another point. In addition, it is of course useful to assign a name that makes it clear what can trigger the optional use case.

The new use case *Select file format* must now be connected to the main use case with a directed connection. The representation is identical to the <<*include*> > −connection, but points from the extension to the main use case and is marked with the <<*include*> > note.

In addition, a comment field is inserted that is attached to the <<*include*> > −connection with a dashed line. The comment field specifies with condition: the circumstances under which the extension is executed and defines the extension point to which the extension belongs.

With very complex software systems, it is easy to lose track of what is going on and it is very important to break the software down into components that are responsible for specific tasks.

In this case, it is of course also important in the use cases that the components affected by the various use cases can be documented. The UML allows the documentation of responsible software components in use case diagrams by drawing simple rectangles around the affected use cases.

Figure 2.8 shows the already known example in a simplified form and supplemented by the Responsible Software Component.

Within the rectangle, all use cases concerning the software component can be documented. The name of the software component must be documented in the upper left corner. The keyword <<*Subsystem*> > makes it clear that it is a software component.

The use case diagrams of UML can actually do much more than has been described so far, but this description should suffice for a start. A complete documentation can be found in the respective current UML specification (Object Management Group 2018).

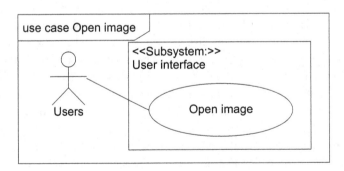

Fig. 2.8 Use case diagram for loading an image and assigning the responsible component

2.2.2 The Activity Diagram

The use case diagrams serve as a first very rough estimate of the tasks of the software system. And even if the use case diagrams already offer a variety of possibilities, they are still not sufficient to describe functional processes in detail.

For each use case, many small steps must be performed in a specific sequence, which depend on conditions, partly run in parallel or open up different choices. To be able to describe this in detail, activity diagrams are needed. To describe some basic elements first, Fig. 2.9 shows a very abstract activity diagram for the use case *Open image*.

In the upper left corner, the name for the diagram is defined. The abbreviation *act,* or the word *activity*, makes it clear that this is an activity diagram. This type of diagram always requires a starting point at which processing begins. One way to do this is to have a starting node, which is represented as a completely filled black circle. Starting from this node a *token* travels along the directed connection, which is called edge, or *activity edge.*

A *token* is a marker or data container that moves through the network. If the *token* encounters an activity node, this node is activated and executed. All other nodes are inactive. In this first example diagram, the activity node was called *Open Image* and was represented by a rectangle with rounded corners. What exactly happens inside this activity node can be shown in a later diagram.

After the activity *Open Image is* completed, the *token* follows the edge again and meets the end node, which ends the parent activity.

The start and end nodes belong to the so-called control nodes, which serve to control the movement of the *tokens* within the diagram. Activity nodes, like *open image* are called executable nodes.

The activity diagram in Fig. 2.10 is intended to specify in more detail what happens within the activity *Open image.* In order to make it clear that we are dealing with the processes within a node, a further frame has been inserted to represent the boundaries of the activity node. Additionally, some new display elements have to be introduced.

First, since its version 2.0, UML distinguishes between control flows and object flows. The control flows determine when activity nodes are executed. The *tokens of* the control flows originate in the start nodes and end in the end nodes. On their way they activate the activity nodes, are redirected by decision nodes, or possibly even sent to two parallel paths by a fork.

In addition, however, UML allows object flows to define which data is exchanged between different activity nodes. For this purpose, the activity nodes are extended by small rectangles, the so-called *pins,* each of which stands for a parameter that either leaves the

Fig. 2.9 Activity diagram for loading an image

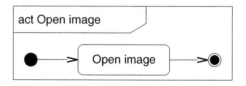

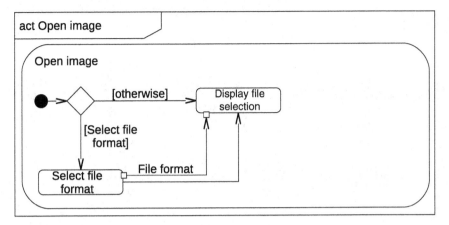

Fig. 2.10   Activity diagram for the processes within the activity *Open image* (part 1)

node or reaches it. In the diagram, object flows can always be recognized by the fact that they begin and end at *pins*, while control flows are connected directly to the activity node.

Decision nodes are represented by a diamond. These nodes allow to redirect the control flow under certain conditions. These conditions, called *guards*, must be written inside square brackets on the edges leaving the decision node.

In the following, a distinction is made between *object tokens* for object flows and *control tokens* for control flows.

After the activity node *Open image* has been activated, a *control token is* created in the start node that follows the edge to the decision node. If a selection of the file format is desired, the *control token* follows the corresponding edge. Within the activity node *Select file format*, an *object token is* generated that determines the file format. This leaves the activity node via the corresponding pin and is transferred to the input pin of the activity node *Display file selection*. When the activity node Select file format has finished its activity, the *control token* moves over the edge to the activity node *Display file selection* and activates it.

Alternatively, the activity node *Display file selection can be* activated directly if no selection of the file format is desired. In this case, no *object token* is transferred to the node. This empty or unset object is also called *null*.

What happens inside the node *Display file selection is* described in more detail in Fig. 2.13. First, however, the further process after the file selection has been completed is to be considered. For this purpose, the diagram in Fig. 2.10 is supplemented so that the diagram shown in Fig. 2.11 results.

Two new *pins have* been added to the activity node so that it can generate two *object tokens.* One of the two *object tokens* transports the selected file name to the next activity node *Load image.* However, the second *object token* is much more exciting. It transports the information whether the selected file should actually be opened and has either the value "*true*" *(true)* or "false" (false). This information is forwarded to a decision node.

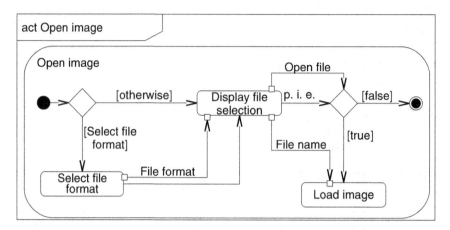

Fig. 2.11 Activity diagram for the processes within the activity *Open image* (part 2)

When the activity node *Display file selection* has completed its tasks, a *control token* also moves to the decision node. Thus, an *object token* and a *control token* now move to the decision node. The edge of the control flow has been labeled with the three letters *p. i. e.* This stands for *primary incoming edge* and states that the decision node forwards only the control flow. However, the object flow is used to make the decision and is called *decision input flow.*

If the selected file is actually to be opened, the *control token is* forwarded to the *Load Image* activity node. Otherwise, the activity is canceled.

Figure 2.12 now shows the complete sequence of the activity *Open image.*

Load image also creates two *object tokens.* One of them transports the loaded image to the activity node *Display image.* The second *object token* is transported again to a decision node that decides whether the image should be displayed or not. The *object token* contains information about whether the image was loaded successfully. However, this can fail for various reasons. Perhaps an external drive has been removed in the meantime and the file can no longer be read. It is also possible that the selected file was not an image at all. In these cases, of course, the *Load Image* activity node could not complete its task successfully and there is no image to display.

The construction of the decision node is analogous to the previous example. The node has two inputs, each carrying a *control token* and an *object token.* Again, the control flow is the one that is forwarded, while the object flow is only used for decision making.

It is often useful to document the tasks of individual activity nodes even more precisely. In the example, this certainly applies to each activity node shown in Fig. 2.12. For the introduction of the activity diagrams, however, only one further activity node is to be documented in more detail. Figure 2.13 shows the processes within the node *Display file selection.*

The edge of the activity has again been shown within the diagram to make it clear that these are processes within a node. In addition, however, there are rectangular nodes on the

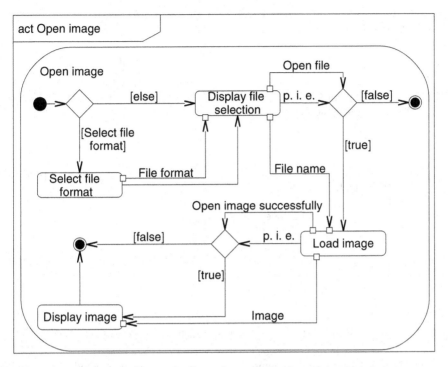

Fig. 2.12 Complete activity diagram for the processes within the activity *Open image*

edge of the activity. These are called object nodes and represent the *pins from* Fig. 2.12. They are therefore the inputs and outputs of the activity's object flows.

Directly after the input object node *File Format,* the *object token* reaches a decision node. Unlike all other decision nodes described so far, this decision node is located on an object flow. Its behavior is not changed by this, but this decision node can only redirect *object tokens.* In this case, it distinguishes between two cases. If no *object token* was passed to the activity, the file format is *null.* The *object token is* redirected to the object node *Default file format* and transmitted to the activity node *Filter file selection.*

In this representation, the activity node has no *pins.* This notation is permitted in UML if an object flow runs explicitly through an object node. In this case, the node takes over the task of the *pin.*

If a file format was passed to the activity node *Display file selection,* this *object token is* passed to *Filter file selection.*

The control flow of the activity starts at the start node as soon as the activity is activated in the parent diagram. The *control token* first reaches the activity node *Filter file selection.* This activity generates an *object token* with the selected file, which is transmitted to the output pin.

After completion of the activity, the *control token* moves on to the decision node, which checks whether the file should be opened or not. In the first case, the *object token moves to*

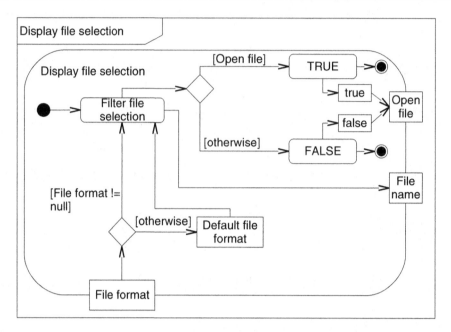

Fig. 2.13 Activity diagram for the processes within the activity *Display file selection*

the action *TRUE*. An activity that cannot be shown in more detail is called an action. The action *TRUE* generates the *object token true* and forwards it to the output pin *Open file*. The control flow is terminated after the action is exited.

The action *FALSE* behaves similarly, except that an *object token false* is passed to the output pin.

To fully describe the activity, all activities would need to be represented by diagrams.

For activity diagrams, the statement about use case diagrams applies even more. The UML offers many more elements and notations that extend the possibilities of activity diagrams. The representation in this book is only a small part of it, but it already allows the representation of complex processes. Again, full documentation can be found in the latest UML specification (Object Management Group 2018).

However, three important elements of activity diagrams that have not been mentioned so far will be added in the following two chapters.

Connectors

When describing complex processes, the many different control and data flows can easily form a confusing network. As soon as it is no longer possible to represent the connections without intersections, the comprehensibility of the represented information suffers additionally. The UML therefore allows the use of so-called connectors, which are shown in Fig. 2.14.

Fig. 2.14 Connectors in an activity diagram

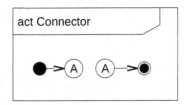

Fig. 2.15 *Fork* and *Join* in an Activity Diagram

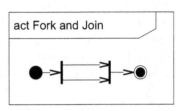

A connector is represented by a circle containing an identifier. If a flow is directed into a connector, it can reappear at exactly any point in the diagram by again inserting a connector with the same identifier. For the activity diagram, the connection is considered to be a continuous flow. However, the clarity of the representation is significantly increased.

Fork and Join

Occasionally, the results of object or control flows are required at different points in an activity, because processes are to take place in parallel or objects have to be processed at two points. UML offers the possibility of splitting a flow with the help of a *fork or* merging it again with the help of a *join*. In Fig. 2.15, a control flow is split using a *fork* and then merged again using a *join*. Both are represented by thick black lines that split or join flows.

In a *fork,* a flow can be split into multiple flows without changing the nature of the flow. If a control flow is split, only control flows are created. Object flows behave in the same way.

A *join* ensures without further description that the processing only continues when a token is present on all incoming flows. Incoming flows are therefore synchronized. If there is at least one object flow among the incoming flows, the outgoing flow is always an object flow. The result is a control flow if only control flows lead into the *join node.*

2.2.3 The Class Diagram

Once the processes of the use cases have been described using the activity diagrams, program structures in which these processes take place must emerge at some point. In order to document the various classes and their relationships to one another, class diagrams are used in UML.

This type of diagram describes the static structure of programs developed using object-oriented approaches. The idea of object-oriented programming and the classes in *C++* are described in Chap. 10. Without this knowledge, some of the concepts described here, such as inheritance or abstract classes, are difficult to understand.

In addition, the class diagrams are already very close to the actual implementation, although it is irrelevant with which language the diagrams are later realized. When reading the description, the question therefore often arises as to how a concrete issue can be implemented. However, this is not decisive at first. More important are the considerations which functions and variables are theoretically needed.

Class diagrams are already very close to the actual implementation in the program code. This even goes so far that many UML design programs offer the option to generate class diagrams from already written source code. This is called reverse *engineering* and makes it clear that the diagrams should have been there before. The reverse case is also offered, in that the associated source code is immediately generated from a diagram. Consequently, this process is called *forward engineering.*

Figure 2.16 shows a class diagram that illustrates three different ways in which classes may be represented in diagrams.

A frame is also drawn around this diagram type. The keyword class makes it clear that this is a UML class diagram.

In the simplest case, a class is represented by a simple rectangle. The name of the class is written in bold in the center of the field. This representation can be useful if a large number of classes appear on a diagram and a detailed representation of individual classes would reduce clarity.

At a very early stage of software development, however, it can also make sense to initially document all tasks in the form of class names without specifying them in more detail.

The second class representation already offers more information. A second rectangle lists the variables that belong to the class. In UML, these are called *attributes*. First, the name of an attribute is written down, followed by a colon and then the respective data type. In a third rectangle, the functions of the class are listed, the so-called *operations*. These are indicated by the two round brackets at the end. Function parameters are inserted within the brackets in the already known notation (name, colon, data type). In this representation, the visibility levels of the attributes and operations are missing.

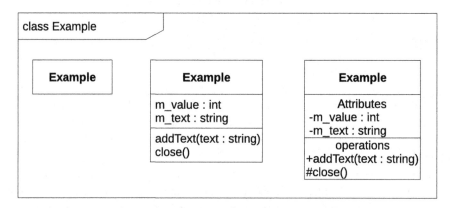

Fig. 2.16 Class diagram for the different display options of a class

These are added in the third display variant. The plus sign (+) stands for the visibility level *public*,[1] the hash sign (#) for the visibility level *protected*[2] and the minus sign (−) for the visibility level *private*.[3] In this display, the rectangle for the attributes is given the heading *attributes* and the rectangle for the operations is given the heading *operations*.

The third representation is particularly suitable if a part of the software is to be described in great detail, since all attributes and operations are represented. Connections between classes can also be easily traced if it is already apparent from the attributes that the data type of an attribute corresponds to another class.

Now the example *Open image* is to be concretized by a class diagram. Figure 2.17 shows the classes with which the task is to be solved.

The class *Frontend* shall implement all tasks that involve a direct interaction with the user, while the class *Image is* responsible for all image operations. Finally, the display of image data shall be done by the class *View*. This structure has proven itself for data processing software, which is why it belongs to the standard designs of software architecture.

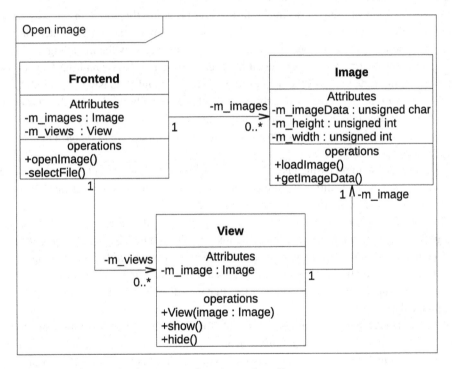

Fig. 2.17 Class diagram for the use case *Open image* (part 1)

---

[1] Full access to the elements from inside and outside the class.

[2] Full access to the elements from inside the class. Access from outside the class is prevented.

[3] Full access to the elements from within the class. Access from outside is not allowed, as with *protected*. In addition, the element cannot be inherited. The term is explained again in detail in connection with the classes.

The technical term for such standard designs is *design* pattern. This particular design pattern is called a *model view controller*.

The role of the *controller is* taken over by the class *Frontend*. It controls the processes between the different classes. In this example, the class stores a list of images in the attribute *m_images* and a list of views in the attribute *m_views*.

Since the data type of the attribute *m_images* is the class *Image*, which is also shown in the diagram, a directed *association* between the class *Frontend* and the class *Image* was inserted into the diagram. The *association* is represented graphically by a solid line with an arrowhead. The direction indicates that the class *Frontend has* a reference to the class *Image,* but not vice versa.

At the end of the *association is* a label *−m_images* which makes it clear that the relationship was established by this attribute.

The numbers below the *association* are called multiplicities and mean in this case that an object of the class *Frontend* can be connected to any number of objects of the class *Image*. Here, means that $0..^*$ also no connection is allowed.

From the point of view of an image processing program, this modeling makes sense insofar as the program certainly has not yet loaded any images at startup. Later, during operation, no maximum number of images is given to the users by the modeling.

A similar association has been added between the *Frontend* class and the *View* class. This allows multiple views of the images to be managed simultaneously.

The class *Frontend* must, as already mentioned, additionally implement the tasks that require direct user interaction. In this example, these tasks are handled by the operations *openImage* and *selectFile, which* correspond to the activity nodes *Open Image* and *Display File Selection from* Fig. 2.12.

The next steps of processing take place within the *Image* class. This class is responsible for loading, saving and editing the image data. To do this, it needs attributes that can store all the necessary information about the stored image.

In one of the simplest cases, this is the image data itself, stored in the attribute *m_imageData,* and the width and height of the image, stored in *m_width* and *m_height*.

The first operation *loadImage* implements the activity node *Load Image* from Fig. 2.12. For the second operation *getImageData* there is no equivalent in the activity diagrams yet. However, it is necessary that an external class that is to display the image can somehow get the image data.

The third class in the diagram is responsible for displaying images and thus fulfills the tasks of the activity node *Display image* from Fig. 2.12. The class requires an attribute *m_image in* which the image to be displayed is stored. This connection between the two classes is again represented by an additional *association*. The multiplicities at the ends of the *association* make it clear that an object of the class *Image* is uniquely assigned to an object of the class *View*.

The operations *show* and *hide are* used to open and close the image display. The *View* constructor passes the image object to be displayed to the *View* class.

With these three classes, the use case *Open image* can be realized in a software system.

However, two additional display elements are needed to complete the basics of class diagrams. For this reason, the use case is to be extended a little.

In a real image processing software, images are usually displayed in windows of the respective operating system. These windows are usually realized by classes in the software. However, as the images are edited, the displayed contents change occasionally and the display must be adapted.

Of course, each image can now be assigned the class in which it is displayed, but this solution is not very flexible. If the image can be displayed in several different classes, this solution quickly becomes confusing. But even for this case there is a design model that solves the problem, the so-called *Observer Pattern*.

To do this, an abstract class is defined that acts as an independent interface for all classes that need to be updated. Updating occurs whenever the data of a model, in this case the image, changes.

The screen can now keep a list of interfaces without worrying about what classes might be behind them. Every time the data changes, the list is passed through to inform about the update.

Figure 2.18 extends the already known class diagram by an *Observer Pattern*, with the help of which the class *View* can be informed about every change in the data of the class *Image*.

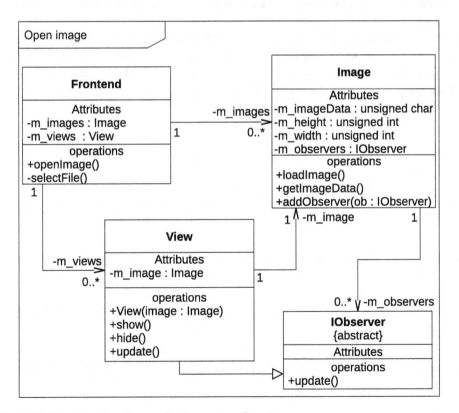

Fig. 2.18 Complete class diagram for the use case *Open image*

To do this, we first introduce an abstract class (see Sect. 10.5) *IObserver, which* has only one operation *update()*. The capital "I" in front of the name of the class makes it clear that this is the definition of an *interface*. However, this naming is only a convention of good practice to make interfaces easier to identify within the later program.

To ensure that the class is actually defined as an abstract class within the UML, it is necessary to note the keyword *abstract* in curly brackets under the name of the class.

An attribute is now added to the *Image* class in which all interfaces that are to be informed about an update can be stored. In this example, the attribute was named *m_observers* and has the data type *IObserver*. In addition to the attribute, an *association* was added that refers from the class *Image* to the class *IObserver*. The multiplicities determine that a connection to several interfaces is possible.

In addition, an operation *addObserver* must be added, which can be used to add a new observer to the list.

Now the *View* class is to inherit from the *IObserver* class so that it can be added as an interface to the Observer list of the *Image* class. This is indicated in UML by a directed connection called *generalization*. This connection is represented by a solid line with a framed arrowhead at the end. The arrowhead points from the derived class to the base class, i.e. from the specialization to the generalization.

Finally, the *update* operation is added to the *View* class to implement the inherited interface.

Once again, many representation elements of UML have not been mentioned, yet a very complex example has already been created. The complete documentation of a UML class diagram can be found in the respective current UML specification (Object Management Group 2018).

Taxonomy Levels

3

The exercises in this book are intended to enable you to assess your own abilities. This requires that you can decide for yourself whether you have solved a task or not. In addition, you should receive feedback about which skills are tested with the respective tasks.

In 1956, the American psychologist Benjamin Bloom developed a system of six cognitive taxonomy levels that can be used to categorize learning objectives (Bloom et al. 1956).

The first level is *knowledge* and describes the ability to remember and reproduce what has been learned. At the second level, *comprehension,* facts can be described and explained. The third level, *application, is* about using what has been learned to solve concrete problems. The fourth level is *analysis*. Here it is a matter of distinguishing and comparing different solutions. At the fifth level, *Synthesis,* several solutions can be combined to a common solution. And in the last stage, the *evaluation,* own solutions can be developed and existing solutions can be evaluated.

Gerwald Lichtenberg and Oliver Reis highlight in Lichtenberg and Reis 2016 that in addition to the categorization into taxonomy levels, a further scale is necessary for the assessment. These levels can be defined individually for each taxonomy level. However, this procedure requires that another person exists who can assess the responses.

Since you should be able to use this book to assess your own learning success, Lichtenberg and Reis have simplified the procedure somewhat.

Each task in this book has been assigned a taxonomy level, which you can identify by a pyramid located above each task. Table 3.1 shows the different taxonomy levels and how they are marked in the exercises. In addition, the table shows you which criteria you can use to assess whether you have solved the task or not.

For each correct solution you can give yourself one point in the respective taxonomy level for a chapter.

Table 3.1 The taxonomy levels for cognitive learning objectives according to Bloom

Description	Symbol
Knowledge All terms correctly used and coherently reproduced	
Comprehension The task explained with a proper rationale	
Application Calculating the correct result and applying the correct solution method	
Analysis Creating a correct hypothesis	
Synthesis Create a feasible proposal. If necessary, provide proof by testing	
Evaluation Development of a target-oriented solution idea	

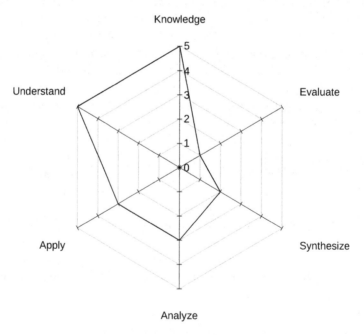

Fig. 3.1 Network diagram for self-assessment using taxonomy levels

In the exercises, you will find a network diagram for each chapter, as shown in Fig. 3.1. The black line shows you the maximum number of points you can get in a taxonomy level in this chapter.

After you have worked through the tasks and scored them for yourself, you can compare your score with the chart. Your scores for the respective taxonomy levels show you in which areas you were able to collect points. With the assessment criteria from Table 3.1, your results thus correspond approximately to level 3 according to Lichtenberg and Reis (2016).

Part II

Basics

The Language

4

This chapter introduces all the basic language elements of *C++*. It is intended both as a reference work for the problems presented in Part III and as an introduction to the programming language. For this purpose, the individual language elements are introduced with the help of small examples. The acquired knowledge can be checked with the help of the exercises in the self-test. As explained in the previous part of the book, the tasks are marked with taxonomy levels so that you can better assess your own abilities.

The *C++* language is an extension of the *C* language. The *C* language was developed in the early 1970s and follows a procedural programming paradigm. At the end of the 1970s, the *C* language was extended by Bjarne Stroustroup with the so-called object-oriented programming paradigm. What object orientation means exactly is explained in Chap. 10 about classes. However, the first chapters of this part first deal with the procedural development of programs.

Various commercial or non-commercial development environments are available for developing *C++* programs. Each of these development environments occasionally changes the way the *C++* programs are to be written, and sometimes their appearance. Some support you by already generating program frames through the development environment, while others do not. To keep this book as independent as possible of the particular development environment, no particular development environment is assumed. Of course, this has the disadvantage that the programs presented will only run with slight modifications in certain development environments.

The programming language *C++* is a so-called *compiler language*. This means that a program, more precisely the source code, is translated by another program, the *compiler,* into a language that can be understood by the processor on which the program is to run. The source code can consist of several files. The *header files* contain only abstract information, comparable to the table of contents of a book. The *cpp files*, on the other hand,

B. Tolg, *Computer science to the Point*,
https://doi.org/10.1007/978-3-658-38443-2_4

contain the complete definition of the program elements, practically the book chapters that match the table of contents.

In *C++,* the translation takes place in three steps. First, the *preprocessor makes* modifications to the source code that enable the *compiler to* work, but would make it more difficult for the programmer to read the source code. In the second step, the now modified source code is translated by a *compiler.* The result of this translation are object files, which in principle could already be understood by the processor. However, some programming elements extend over several files, so that open ends (so-called *links*) arise. These open ends still have to be connected, this is done in the last step by the *linker.* The result of these steps is either an executable program or a library. The latter also consists of executable code, but has no main function where the program could start. You will learn what a main function is in the following Sect. 4.1.

4.1   The First Program

Following an old tradition in learning programming languages, the first program a student of a new programming language should write is a *Hello World!*-program. The program does nothing more than print the aforementioned script to the console. Listing 4.1 shows the corresponding source code.

Listing 4.1 The *Hello World!* Program

```
 1   # include <iostream >
 2
 3   using namespace std;
 4
 5   int main ()
 6   {
 7     cout << " Hello World !" << endl ;
 8
 9     cin .get ();
10
11     return 0;
12   }
```

Much of what is explained in this section appears again elsewhere and usually in more detail. Nevertheless, each command of the program is to be examined roughly at first, in order to enable an introduction to the language *C++.*

The program starts with the preprocessor command, or the *preprocessor directive* *#include*. When editing the source code, the preprocessor recognizes the *#include* command and replaces the line of source code with the complete contents of the file *iostream*. The filename following the *#include* command may be enclosed in angle brackets

(*#include < filename>*) or in quotes (*#include "filename"*). The difference between the two notations is the number of directories that are searched to find the file.

If angle brackets are used, only those directories are searched which are passed to the *compiler* by the development environment when it is called or, if the search is unsuccessful, which are specified by a certain environment variable. This usually does not include the directories where your own project is located. The quotes specify a different search order. Here, the directory where the file containing the *#include is* located is searched first. This is followed by the directories of all other files opened by *#include,* and only then the directories that are also searched with the angle brackets.

The quotation marks thus specify a larger search radius. In the examples, angle brackets are nevertheless always used when possible.

The *iostream* file defines the *cout, cin,* and *endl* statements, which could not be used without the *#include.*

The third line of the program contains the command *using namespace std;*. A *namespace* can be created around certain elements of the *C++* language. In principle, it works similar to a family name. In this case, the namespace *std* has been defined in the file *iostream* around the statements *cout, cin,* and *endl.* Therefore, if the *cout* command is to be called, the *std* namespace would always have to be called first. The call would be *std::cout.* Since this can become unreadable with many namespaces, *C++* allows to do without naming the *namespace* if this was announced beforehand by the *using namespace* command.

In general, all statements in *C++* must be terminated by a semicolon. However, as is almost always the case, there are exceptions to this rule. For example, preprocessor statements such as *#include do* not require a semicolon, and this is not always the case after curly braces.

In line five of the program follows the line *int main()* and then some commands enclosed by curly braces. As mentioned before, the *C* language and also its extension *C++* support a procedural programming paradigm. In this case, this means that programs can consist of several parts called procedures, or functions. These parts perform certain tasks and are given a name. This name can be chosen arbitrarily within the limits described in the syntax diagram 2.3 on page 6. However, it is advisable to use the name to describe the function's task in summary form. This simplifies the work with the program considerably.

A function can, as may be known from mathematics, receive values and also return them in various ways. The line *int main()* causes the definition of such a function. This is the most important function of the program, the so-called main function. It is the entry point to your *C++* program and every program must have such a function somewhere. No matter where it is located within your source code, the program always starts there.

The definition consists of four important areas. First comes the *int.* This is a variable type that can store positive and negative integers. In this case, it is used to define the return value of the function. When this function is called, it is clear to the caller that the result of the call will be an integer.

The second area is the *main.* This is the name of the function, which can normally be freely chosen within the described limits (syntax diagram 2.3). In the case of the main function, however, this name is subject to additional restrictions. In order for the main function to be recognized as such, the name *main* must be chosen.

In third place come the brackets *()*, which in this example remain empty for the time being. Nevertheless, they are very important for the definition, because they make it clear that a function is to be defined. In later examples, the values to be passed to the function, the so-called *parameter list*, will be inside these brackets. From mathematics, you may be familiar with the $y = f(x)$ typical function definition, which works on the same principle. Inside the parentheses is also in mathematics the parameter that is to be passed.

Finally comes the area inside the curly braces in lines six through twelve. The curly braces mark the function body. This is the area that defines what is to happen when the function is called. Program statements can only be inside functions because that is the only way a defined program flow can occur. Instructions outside of a function could not be processed meaningfully, since it would not be clear when they are to be executed.

Within the function, three commands are executed. First *cout ≪ "Hello World!" ≪ endl;*.

The command *cout* is defined, as already mentioned, in the file *iostream*, which was inserted in the first line by *#include*. It causes an output in a text console. The individual outputs can now be written after the *cout*, separatedby ≪ from each other. In the example, the text *Hello World!* is to be printed. To mark the output as text, quotation marks must be placed around the text (*"Hello World!"*). Finally, you want to jump to the next line. In word processing software, this would be done by pressing the *Enter key*. In the program this is done by the command *endl*.

What does the program do so far? It starts in the main function and prints a text on the screen. Thus it has fulfilled its desired purpose. Now the program could terminate itself directly, but this would happen so fast that nobody could read the text. So it is necessary to wait for an interaction of the user before the program is terminated. This purpose is served by line nine of the program *cin.get();*, which waits for the user to press the *Enter key*.

The command *cin*, like *cout*, comes from the file *iostream*. However, the command does not write a value to the text console, but reads a value from there. The round brackets of the *get()* function make it clear that it is a function call, which again takes no parameters. The dot between both commands shows that the function *get()* is provided by *cin* and could not be called alone.

Line nine is not required in every case and every development environment. Some development environments leave the text console open after the program has finished, so that the result remains visible. In the sample programs in this book, the command has been omitted. If your development environment closes the console immediately after execution, you only need to write the line *cin.get();* directly before the *return in* the sample programs.

In line 11 now follows the last command of the program *return 0;*. This command terminates the function and returns the value 0 to the caller. In the function definition in line 5, the main function was defined by the *int to* return an integer. This definition is now satisfied by the 0.

Again, the main function is special because it is called by the operating system when the program starts. The operating system expects the return of the value 0 in case the program was executed properly. Any other value is considered an error, even if the operating system cannot interpret this value. When developing a program, care should therefore be taken to ensure that there is documentation of the possible return values so that error codes can be interpreted.

4.2   A Few More Tips

First, check whether your development environment automatically closes the text console after executing the program. If this is the case, you must add the line *cin.get();* before the line *return 0; in* all the sample programs in this book.

You may have noticed that the source code in Listing 4.1 follows a certain structure. Curly braces, for example, always stand alone on a line and the closing brace always stands exactly below the opening one, thus enclosing a block. Within a block, all lines are indented so that they are also placed one above the other.

This is by no means assumed by C++. Listing 4.2 works the same way as Listing 4.1, and you can judge the readability for yourself.

Listing 4.2 The *Hello World!* Program, Different

```
1   # include <iostream >
2   using namespace std;int main ()
3   {cout <<" Hello World !"<< endl ;cin.get (); return 0;}
```

It is wise to follow certain conventions from the beginning. In this book, certain rules are followed consistently and I will refer to them at the appropriate time. There are certainly different points of view on the details of the conventions, and you may eventually develop your own rules. For starters, however, I recommend that you try to imitate the style of the program examples shown here.

Use spaces, tabs and new lines to increase the readability of your programs. For the *compiler,* these characters, called *whitespaces, are in* most cases unimportant. However, they are indispensable for the readability and maintainability of programs.

This also includes adding comments to your own programs. In C++ there are two possible types of comments.

- Single-line comments started by *//.* The rest of the line is now interpreted as a comment and is no longer subject to C++ language rules.
- Multiline comments started by the string */** and ended by **/.*

Listing 4.3 shows the familiar *Hello World!* program with comments.

Listing 4.3 The *Hello World!* Program with Comments

```
 1   /*
 2   Author : Boris Tolg and many before and after him.
 3   Title : Hello World !
 4   Date : 02.03. - - - -
 5   Language : C++
 6   ...
 7   */
 8
 9   # include <iostream >
10
11   using namespace std;
12
13   int main () // Main program
14   {
15     cout << " Hello World !" << endl ; // Text output
16
17     cin .get (); // Waiting for user input
18
19     return 0; // End of the program , Status ok
20   }
```

You should get into the habit of commenting your own programs in detail from the beginning. You should take care that the comments should give an abstract description of the contents. A comment *alphabetical sorting of the words* for several instructions in a small program section facilitates the understanding of a program much more than a comment *Here the value 0 is returned* before the line *return 0;*.

In addition, you should frequently have your programs compiled by the development environment. This has two main advantages:

- You will recognize very quickly if you have written something wrong in your program.
- Since the error messages of the *compiler* or *linker* are often not very clear, the area you have to search for the error is manageably small.

Variables

5

Short and Sweet

- Variables allow you to store values within programs.
- There are different types of variables.
- Variables of a certain type have a fixed value range.
- A variable type also specifies the type of values that can be stored:
  - Integers,
  - Comma numbers,
  - Truth values,
  - Letters.
- **Possible sources of error:**

When converting values of a certain type to another, content may be lost.

Variables are one of the most important elements of any programming language. They are used to store values, to pass them on or to assign meaningful names to certain values. In Listing 5.1, various variables are defined and initialized.

Listing 5.1 Variable Definition and Initialization

```
1   // global variables
2   int a;      // Definition
3   int b = 10; // Definition and initialization
4
5   int main() // Main program
6   {
7     // local variables
```

© The Author(s), under exclusive license to Springer Fachmedien Wiesbaden
GmbH, part of Springer Nature 2023
B. Tolg, *Computer science to the Point*,
https://doi.org/10.1007/978-3-658-38443-2_5

```
8      int x; // Definition
9      int y = 0; // Definition and initialization
10   }
```

Before a variable can be used in *C++*, it must first be defined somewhere in the program. Two pieces of information are needed for this:

- The type of the variable (e.g. *int*), which determines how much memory is required for the variable and which values are to be stored in it.
- The name with which the variable is to be addressed in the future. Names may contain letters, numbers and underscores, but may not begin with a number.

In line two, a variable with the already known variable type *int* for integers and the name *a is* defined by *int a;*. In addition to the definition, the variable can also be initialized immediately. This is done in line three by *int b = 10;*. The variable *b* is not only defined as an integer, but also initialized with the value 10. Variables that have not been initialized have a random content. Therefore, get into the habit of always initializing variables. This will help you to avoid errors, especially at the beginning.

Now the position of the definition or initialization decides about the scope of the variable. If a variable is defined outside of a function, it is considered a *global* variable. This means that it is known and can be accessed in every function of the program.

If the variable is defined or initialized within a function, it is referred to as a *local* variable that can only be used within the function. In addition, the lifetime of the variable is also dependent on that of the associated function. It is created when the function is called and destroyed when the function terminates. If *local* variables use the same name as *global* variables, a confusing situation arises: For the duration of the function call, the name is used to address the *local variable*. During the function, this variable can be read and changed. When the function ends, the *local variable is* deleted again and the *global* variable with the same name remains. Its value has not changed.

If such a situation occurs, it makes sense to search the immediate vicinity to find all definitions for variables of that name. Normally, the closest variable definition will then be responsible for the variable you are looking for.

This makes working with *global* variables problematic. Also the memory usage is mostly not optimal, because the variables always exist. Much more effective are *local variables, which* only exist when they are needed. A final example is parallel function calls, which can cause major difficulties with *global* variables.

All this has led to *global* variables being considered a bad programming style. There is practically no case where a better solution could not be found. The best thing to do is to get into the habit of doing without *global* variables right away.

5.1 Variable Types

The C++ language, like many other programming languages, distinguishes between integer and real data types. The amount of memory used determines how large the value range is that can be mapped by the respective variable type. Table 5.1 gives an overview of the most important data types of the C++ language.

As you can see from the second column of the table, many variable sizes are not specified by the C++ language standard and may vary depending on the *compiler* used. The third column shows the values used in this book.

- The variable type *void* is used as a placeholder. It does not express a special variable type and has no defined size. For this reason, no variables of type *void* can be defined. Nevertheless, it is of great importance for functions and pointers and will be described in more detail in Chaps. 8 and 9.
- The variable type *bool* can be used to store the results of logical expressions. They can either take the value *true*, or *false*. They become relevant in Chaps. 6 and 9.
- Variables of the type *char* are used to store individual letters or text characters. The letters are assigned to numbers. The assignment is based on the so-called *American Standard Code for Information Interchange,* ASCII for short. To mark single text characters in C++, single quotation marks are needed. For example, if the letter *a* is to be assigned to the variable *letter*, this is done in C++ by initializing *char letter = 'a';*.
- The variable types *short*, *int*, *long* and *long long* can be used to store integers. The number range that can be covered by the respective variable types is larger the more bytes the variable type occupies. The integer data types can hold positive and negative values. However, if only positive values are to be represented, the keyword *unsigned* can be prefixed to the variable definition. In this case, twice as many positive-only values are available. This is explained in more detail in Sect. 5.5 on number systems.

Table 5.1 Variable types of the C++ language

Variable type	Size (definition)	Size (assumption) (bytes)
void	–	–
bool	Mostly 1 byte	1
char	1 byte	1
short	At least 2 bytes	2
int	At least 2 bytes	4
long	At least 4 bytes	4
long long	At least 8 bytes	8
float	4 bytes	4
double	8 bytes	8
long double	Mostly 10 bytes	10

- *float*, *double* and *long double* are the names of the real-valued data types. These data types are so-called floating point numbers. Since the memory requirement is also fixed for these variable types, a large number of decimal places can be stored for small numbers. For large numbers, however, the decimal point is shifted further and further to the right so that the large amount can be mapped. This reduces the accuracy in the decimal area. The decimal point therefore flows from small and exact to large and imprecise.

To determine exactly how much memory a variable type or variable occupies, the *sizeof* function can be used. For example, *sizeof(int)* returns the value 4, while *sizeof(data)* returns the size of the variable *data*.

5.2   Type Conversion

Sometimes it may be useful to interpret one variable type as another variable type for a particular operation. For example, if the contents of a *double variable* are copied into an *int variable*, the *compiler* will warn that data loss is imminent because the decimal places will be lost. Through a type conversion, a so-called *typecast,* the data type for this one operation is interpreted differently, so that the *compiler* no longer needs to generate a warning.

An *explicit* type conversion is performed by writing the new data type in round brackets in front of the new expression to be interpreted. Listing 5.2 shows two examples of such a type conversion. First, lines 8 and 9 initialize two variables *v* and *x,* where *v is* of type *int* and is given the value 97, while *x is* defined as *double* and initialized to 3.5.

Listing 5.2 An Example of Type Conversions

```
1    #include <iostream >
2
3    using namespace std;
4
5    int main ()
6    {
7        // Variable definition and initialization
8        int v = 97;
9        double x = 3.5;
10
11       cout << v << " << (char)v << endl;
12       // Output: 97 a
13
14       cout << x << " << (int)x << endl;
15       // Output: 3.5 3
16
17       return 0;
18   }
```

The values of the variables are now printed in lines 11 and 14. The first output is unchanged, so that the stored value is displayed. In line 11, the variable *v* is then converted into a letter by *(char)* in the second output. According to the *ASCII table,* 97 corresponds to a, which is also printed.

In line 14, the variable *x* is converted to an integer by *(int)* so that the decimal part disappears in the output.

In addition to the *explicit* type conversion, there is also an *implicit* type conversion. This type conversion is necessary in many places, but is not obviously visible in the source code. This makes *implicit type conversion a* potential source of errors.

For example, if values of type *int* are written to a variable of type *short,* the type conversion is performed implicitly. However, the number range of a 4-byte *int* variable is much larger than that of a 2-byte *short* variable. It can therefore happen that the numbers are larger than the number range that can be represented by *short.* This is called *overflow.* If the numbers are smaller than the representable range, this is called *underflow.*

In the case of a direct value assignment, the error may still be easy to detect. However, it would also be possible that the error occurs during an arithmetic operation and the problem is not obviously recognizable. However, since there is no program error due to the *implicit* type conversion, the *compiler* will not display an error, but only a warning. The problem only becomes visible when the program behaves strangely in certain situations.

However, *implicit* type conversion also allows many conveniences in the source code. For example, if a variable of type *double* is defined and initialized with the line *double value = 5;,* the value 5 is an integer of type *int.* This value is *implicitly* converted to a value of type *double* by a type conversion. The reverse case is also conceivable: *int value = 5.3;.* Here, too, an *implicit* type conversion takes place. However, the *compiler* will warn that the conversion from *double* to *int* will cause a loss of data. This warning can be prevented by an *explicit* type conversion: *int value = (int)5.3;.*

5.3   Enumerations

When developing programs it is often necessary to store different states. Examples are controls for the current color of a traffic light or an elevator. In such programs, the readability of the program is increased by giving the individual states meaningful names. With the help of the *enumeration* type *enum* it is possible to give names to a number of different states very easily.

Listing 5.3 defines an enumeration for the different floors of a building. This is done using the keyword *enum* followed by a name for the enumeration, in this case *Floor.* Within the curly braces then follows an enumeration of various terms to be defined for the enumeration. The *enum* statement creates a new variable type that can be used in the rest of the program. The new variable type can take on the values defined within the curly braces.

Listing 5.3 An Enumeration for Different Floors

```
1    #include <iostream >
2
3    using namespace std;
4
5    // Definition of a new enumeration type
6    enum Floor
7    {
8      BASEMENT ,
9      FIRST FLOOR ,
10     SECONDFLOOR ,
11     TOPFLOOR
12   };
13
14   int main ()
15   {
16     // Initialization of an enumeration variable
17     Floor elevator = BASEMENT;
18
19     // Output of the variable content
20     cout << elevator << endl;
21
22     return 0;
23   }
```

Within the main program, a variable of type *Floor* can now be created and initialized. In this example it is called *elevator* and initialized with the value *BASEMENT*. An *enum* always assigns integer values starting with 0 to the various states. The output of the program is therefore simply 0.

It would also be possible to assign a number directly to the *elevator* variable, but this would first have to be converted to a value of type *Floor* by a type conversion. The line reads:

```
elevator = (Floor)1;
```

The variable *elevator* has then assumed the value *FIRSTFLOOR*. However, the goal of an enumeration, namely to increase the readability of the program code, is thereby led ad absurdum.

Enumerations can also be used to specify meaningful labels for the return values of functions. For example, the main function always returns a value with the statement *return 0;* which can provide the user with additional information about the reason for the end of the program. The 0 stands for an error-free program flow.

If different types of errors may occur, it may be useful to assign a range of numbers to each type of error. Input or output errors then lie in the range 10–49, errors in the calculation in the range 50–99, etc. Enumerations therefore make it possible to assign concrete values to the various elements of the enumeration. All subsequent elements are then simply numbered in ascending order.

An enumeration for various error cases might look like this, as shown in Listing 5.4.

Listing 5.4 An Enumeration for Error Codes

```
1    // ...
2    // Definition of a new enumeration type
3    // for error states
4    enum Errors
5    {
6      OK ,
7      ERROR_READFILE = 10,
8      ERROR_WRITEFILE ,
9      ERROR_CALCDATA = 50,
10     ERROR_CALCFFT
11   };
12   // ...
```

In this example, the elements of the enumeration are assigned the following values:

If the statement *return ERROR_WRITEFILE;* were now written in the main function, the program would terminate. It would be printed via the console that the program has terminated with the code 11. If a table with the error codes exists, the user would now know what kind of error occurred.

If further errors are added in the course of development, these can simply be added to the corresponding areas. In this way, they are automatically assigned numbers that lie in the correct number range without the already existing codes being changed. In order for this to work, generous number ranges must of course already be selected during planning so that the limits of the ranges are not reached at some point.

In addition, it would be easier to read the program code, since the statement *return ERROR_WRITEFILE;* is also more meaningful within the program code than the statement *return 11;*.

5.4   Advanced: *const, external* and *static*

In the programming language *C++* there are further keywords, which are necessary in certain situations for the definition of variables.

5.4.1   const

A very simple and obvious keyword by name is *const*. And first of all, its function is indeed simple and obvious. If the keyword is written in front of a variable type during definition or initialization, this variable is defined as a constant. The value of the variable cannot be changed after initialization. For this reason, constants must always be initialized when they are defined.

This is useful if, for example, mathematical constants such as π or e are to be defined, or if certain values are frequently required in your own program and these are to be replaced by a constant to improve maintainability and readability. Listing 5.5 shows the two alternative notations for initializing a constant.

Listing 5.5 Defining and Initializing Constants with the *const* Keyword

```
1   int main ()
2   {
3     const int N = 25; // Definition and
4                       // Initialization of the constant N
5     int const M = 25; // alternative notation
6
7     return 0;
8   }
```

In line 3, the intuitive notation is used, where the keyword *const* is written before the type of the variable. The alternative notation in line 4 seems somewhat unusual at first, but has exactly the same effect. In both notations, the respective variables become constants. However, for later more complicated use cases, the alternative notation in line 4 is much easier to understand. These use cases will be referred to again in later chapters. They concern pointers, which are treated in Chap. 11, functions (Chap. 9) and classes (Chap. 10).

5.4.2   external

Before a variable can be used in the *C++* programming language, it must first be made known to the *compiler.* There are three different ways in which this can be done.

- During the *declaration,* only the name of a new variable is made known to the *compiler.* However, no concrete variable is created, i.e. no memory is provided for it. This must happen elsewhere so that the variable can actually be used. To *declare* a variable, the keyword *external* must be used in *C++*, e.g. *external double x;.*
- The *definition* creates a new variable by both making the name known (i.e., *declaring it*) and reserving the required memory. If a variable is created by, for example, *double x;,* it is a *definition.* The variable can be used directly afterwards.

- In addition to the *definition*, a variable can be assigned an initial value from the beginning. This value assignment when *defining* a variable is called *initialization*. An example of this is *double x = 3.5;*.

Programs can quickly become very complex and extend over several files or libraries. Occasionally it may then be useful to work with a variable that has already been *defined* elsewhere in the program. In this case, only the name would have to be made known so that it can be used.

This *decalaration* is done by the keyword *external*. Programs 5.6 and 5.7 show an example of this use case. In a project consisting of several files, there is a *cpp file* in which a global variable *num_Values* has been defined (Listing 5.6). In the main program, the value of this variable is to be printed. By *declaring* the variable on line 5 in Listing 5.7, the name of the variable is made known. However, no new variable is created; instead, reference is made to the existing one. The value 25 is then printed in line 13 of the program.

Listing 5.6 file1.cpp

```
1    // Definition of the global variables
2    int num_values = 25;
3
4    //...
```

Listing 5.7 main.cpp

```
1    #include <iostream >
2
3    using namespace std;
4
5    external int num_values;
6
7    int main ()
8    {
9      // The value can be output ,
10     // although there is no direct connection
11     // between the files within
12     // of the same project exists.
13
14     cout << num_values;
15
16     return 0;
17   }
```

As mentioned at the beginning of Chap. 5, however, working with *global* variables is problematic. The mechanisms described in this and the next subchapter can be implemented much better in *C++* using classes.

5.4.3   static

The keyword *static* has three major use cases, which are very different. Two of them relate to later chapters, namely functions in Chap. 9 and classes in Chap. 10. These use cases are described there.

The third use case relates to *global* variables, and directly to the previous example in programs 5.6 and 5.7. If *global* variables are defined within a file, as in this example, they are always accessible from other files by the keyword *external*. This may not always make sense and it may be in the interest of the programmer to prevent this.

If line 2 in Listing 5.6 were replaced with the line *static int num_Values = 25;*, the variable *num_Values* would be protected from access from outside the file and the example would no longer work.

With *global* variables, the keyword *static* thus serves to protect against access from other files.

5.5   Advanced: An Introduction to Number Systems

To understand variables, it is important to have an idea of how numbers are represented and processed within the computer. Everyone has certainly heard of the zeros and ones, but how does it work exactly?

The required calculation methods are already taught in elementary school, but mostly only for the decimal number system. For the computer, however, the binary or hexadecimal number system is much more important. The binary number system only knows the digits 1 and 0. On the computer, this corresponds to a bit, which can also only assume the values 0 and 1. If eight bits are combined, this is called a byte. For the number of bytes that make up the different types of variables, see Sect. 5.1.

First of all, we will take a closer look at the decimal number system. It has the number 10 as its base and uses the digits 0 to 9. The digit 0 only stands for a value if it follows one of the other digits. Then it turns ones into tens and tens into hundreds. If it were to precede another digit, this would be an unusual representation, but would not change the value of the number.

If now counting is considered, at first every digit from 0 to 9 is used once at the place for single digits. After that there is an overflow, because the digits have run out. An additional digit is added in front of it and used to represent multiples of the number ten. So the second digit can be read as a digit multiplied by the base to become a tens digit. Again, the ones are incremented with each digit to end up with the tens increased by one. If at some

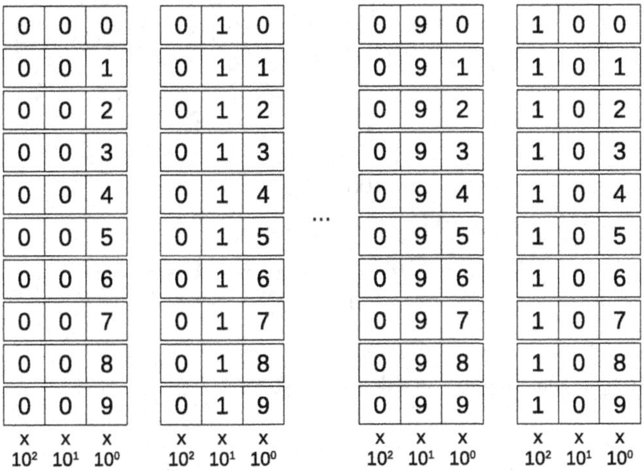

10^2	10^1	10^0		10^2	10^1	10^0			10^2	10^1	10^0		10^2	10^1	10^0
0	0	0		0	1	0			0	9	0		1	0	0
0	0	1		0	1	1			0	9	1		1	0	1
0	0	2		0	1	2			0	9	2		1	0	2
0	0	3		0	1	3			0	9	3		1	0	3
0	0	4		0	1	4			0	9	4		1	0	4
0	0	5		0	1	5	...		0	9	5		1	0	5
0	0	6		0	1	6			0	9	6		1	0	6
0	0	7		0	1	7			0	9	7		1	0	7
0	0	8		0	1	8			0	9	8		1	0	8
0	0	9		0	1	9			0	9	9		1	0	9

Fig. 5.1 Counting digit by digit in the decimal number system

point the tens digits are no longer sufficient, a digit is added to represent hundreds, and so on. The hundreds place corresponds exactly to the digit multiplied by the base squared. A thousand represents the base to the power of three, etc. Figure 5.1 shows the principle.

In the binary number system, counting works according to exactly the same principle, but it uses the base 2. As with the decimal number system, there is no digit for the base itself, so only the digits 0 and 1 are available. Again, the ones are incremented first, but reach the overflow after the digit 1. The binary number system also adds an additional digit, which is also multiplied by the base, which is the two, this time. Figure 5.2 shows the counting principle of the binary numbers.

With this scheme, a binary number can easily be transformed into a decimal number. To do this, simply read the digits from right to left in this representation.[1] The first digit is multiplied by 2^0, then the second digit is multiplied by 2^1 and the values are added, and so on. The result of this sum is the decimal representation of the number. To avoid confusion, numbers are written in curly brackets when calculating with different number systems, and the base of the number system is added. This is usually placed in subscript after the closing parenthesis. Figure 5.3 shows an example of a conversion of a binary number into a decimal number.

Conversely, decimal numbers can be converted into the new number system by a repeated division with the base of the target number system. Decisive here are always the integer remainders that result from the division. Figure 5.4 shows how the transformation

---

[1] The direction of reading depends on the type of number representation on the computer. A distinction is made between *MSB first* and *LSB first*. *MSB* stands for *Most Significant Bit*, i.e. the bit with the highest significance, and *LSB* for *Least Significant Bit*. The *LSB* always stands for the bit that represents the ones digit. This is particularly relevant when numbers are to be exchanged between different computer systems that use a different number representation.

Fig. 5.2 Counting digit by
digit in the binary num-
ber system

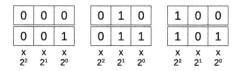

$$\{10011101\}_2$$
$$= \{1 \cdot 2^7\}_{10} + \{1 \cdot 2^4\}_{10} + \{1 \cdot 2^3\}_{10} + \{1 \cdot 2^2\}_{10} + \{1 \cdot 2^0\}_{10}$$
$$= \{128\}_{10} + \{16\}_{10} + \{8\}_{10} + \{4\}_{10} + \{1\}_{10}$$
$$= \{157\}_{10}$$

Fig. 5.3 Converting a binary number to a decimal number

is carried out using the decimal number 42 as an example. First, the number 42 is divided
by 2. The result is 21, and since the division is smooth, the integer remainder is 0. Next,
the result of the division, i.e., 21, is divided by 2 again. The integer result is 10 and since
the division is not smooth, the integer remainder is 1. This procedure is repeated until the
result of the division is 0. The resulting remainders are now the binary number we are
looking for, read from bottom to top.

Why is that? In the first division, the number was divided by 2. The resulting remainder
therefore provides information about whether the number is divisible by 2 as an integer.
The same information is also contained in the last bit of a binary number, because only
with this bit can an odd number be generated. A new division then results in divisibility
by 4, etc.

5.5.1   Addition and Multiplication

The addition and multiplication of binary numbers can be carried out according to exactly
the same principle as was already taught in primary school. The only differences are the
smaller range of available digits and the resulting changes in the rules for carrying over.

Thus the binary addition of the numbers $\{1\}_2$ and $\{1\}_2$ already results in a carryover of
$\{1\}_2$, since the binary representation of the number $\{2\}_{10}$ is $\{10\}_2$. The same applies to a
result of because of the binary representation $\{11\}_2$. Figure 5.5 shows the binary addition
using the example of the numbers $\{157\}_{10}$ and $\{46\}_{10}$.

The rules of multiplication also correspond to those of written multiplication of whole
numbers from primary school. Two numbers are multiplied by multiplying the digits of the
left number individually by the digits of the right number and moving them to the corresponding place. Finally, all the numbers are added up. Figure 5.6 shows multiplication
using the numbers $\{10\}_{10}$ and $\{13\}_{10}$.

Again, when adding up, it should be noted that the binary number system makes it
easier for carryovers to occur.

Fig. 5.4 Converting a decimal
number into a binary number

$42 : 2 = 21$ Remainder: 0
$21 : 2 = 10$ Remainder: 1
$10 : 2 = 5$ Remainder: 0
$5 : 2 = 2$ Remainder: 1 $\{42\}_{10} = \{00101010\}_2$
$2 : 2 = 1$ Remainder: 0
$1 : 2 = 0$ Remainder: 1

Fig. 5.5 Addition of
binary numbers

$$
\begin{array}{c}
\text{Binary} \qquad\qquad \text{Decimal} \\[4pt]
\begin{array}{c}
1\ 0\ 0\ 1\ 1\ 1\ 0\ 1 \\
+\ 0_0\ 0_1\ 1_1\ 0_1\ 1_1\ 1_0\ 1_0\ 0 \\
\hline
1\ 1\ 0\ 0\ 1\ 0\ 1\ 1
\end{array}
\qquad
\begin{array}{c}
1\ 5\ 7 \\
+\ 0_1\ 4_1\ 6 \\
\hline
2\ 0\ 3
\end{array}
\end{array}
$$

5.5.2 Subtraction

To be able to subtract two binary numbers, it makes sense to first develop a representation for negative numbers. On the one hand, the subtraction can then ideally be interpreted by an addition with a negative number. On the other hand, negative numbers are needed to be able to represent the results of the subtraction. It should also be noted that the size of a number on a computer is limited by the number of bits used. Consequently, it is necessary to reinterpret the existing number representations and divide them as equally as possible between negative and positive numbers. In order for the mathematical operations to continue to work, an interpretation must be found that does not change the order of the numbers in a ring representation, as in Fig. 5.7.

This is achieved by inverting each bit of a number individually. A 0 thus becomes a 1 and vice versa. Thus, the first half of the numbers is interpreted as positive numbers in each representation, the second half as negative numbers. Numbers that begin with a 1 are thus considered negative, numbers that begin with a 0 are considered positive. This representation of negative binary numbers is called the (B-1) complement, where B stands for the base of the number system. Binary numbers are consequently referred to as the ones complement. Figure 5.9 shows the now changed interpretation of the binary number patterns. It is clear that the order of the numbers on the ring has not been changed. However, the number 0 appears both as a negative and as a positive number.

To prevent this, an additional 1 is added after the inversion of a binary number. This shifts the representation of the negative numbers, so that there is now no longer a negative interpretation of 0. The resulting representation is called B-complement, or two's complement, and is shown in Fig. 5.9c. Figure 5.8 shows the conversion of the number 92 to two's complement. The back conversion of the number is -92 done by exactly the same steps.

In the context of this book it shall be assumed that for the representation of negative numbers the two's complement representation is used. But even so, there are two possible interpretations for the individual bit patterns, namely once as a negative number, or as a large positive number. This circumstance makes it necessary to tell the computer whether

Fig. 5.6 Multiplication of binary numbers

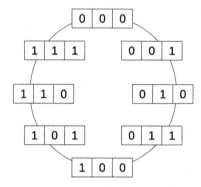

$$
\begin{array}{r}
\text{Binary} \qquad \text{Decimal}\\[2pt]
\underline{1\,0\,1\,0 \cdot 1\,1\,0\,1} \qquad \underline{1\,0 \cdot 1\,3}\\
1\,0\,1\,0 \qquad\qquad 3\,0\\
0\,0\,0\,0 \qquad\qquad 1\,0\\
1\,0\,1\,0 \qquad\qquad \overline{1\,3\,0}\\
\underline{1\,0\,1\,0 }\\
\overline{1\,0\,0\,0\,0\,0\,1\,0}
\end{array}
$$

Fig. 5.7 Ring representation of 3-bit binary numbers

$$
\begin{array}{llll}
\{92\}_{10} & \{\,0\,1\,0\,1\,1\,1\,0\,0\,\}_2 & \{-92\}_{10} & \{\,1\,0\,1\,0\,0\,1\,0\,0\,\}_2\\
\text{invert} & \{\,1\,0\,1\,0\,0\,0\,1\,1\,\}_2 & \text{invert} & \{\,0\,1\,0\,1\,1\,0\,1\,1\,\}_2\\
\underline{+\{1\}_{10}} & \underline{\{\,0\,0\,0\,0\,0\,0\,0\,1\,\}_2} & \underline{+\{1\}_{10}} & \underline{\{\,0\,0\,0\,0\,0\,0\,0\,1\,\}_2}\\
\{-92\}_{10} & \{\,1\,0\,1\,0\,0\,1\,0\,0\,\}_2 & \{92\}_{10} & \{\,0\,1\,0\,1\,1\,1\,0\,0\,\}_2
\end{array}
$$

Fig. 5.8 Conversion of binary numbers in two's complement

negative numbers are to be represented or not. In the variable declaration, it is therefore initially assumed that both positive and negative numbers are to be used. If this is not desired, the keyword *unsigned* can be prepended to the definition of an integer variable. The contents of this variable are then always interpreted as a positive number.

5.5.3 Hexadecimal Numbers

The hexadecimal number system with base 16 works exactly like the already known number systems with base 10 or 2. However, for some applications in computer science it has advantages to use the hexadecimal number representation. The background is that computers are based on the binary number system and many representations are bytes, i.e. combinations of eight bits each. Since the hexadecimal number system has a base of 16, 2^4 bytes can always be represented in hexadecimal by two digits (Fig. 5.9).

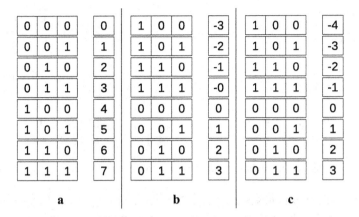

Fig. 5.9 Representation of positive and negative binary numbers (**a**) without consideration of a sign, (**b**) in one's complement and (**c**) in two's complement

Table 5.2 Digit assignment in the hexadecimal number system

Hexadecimal	Decimal
F	15
E	14
D	13
C	12
B	11
A	10
9	9
8	8
7	7
6	6
5	5
4	4
3	3
2	2
1	1
0	0

First, however, to the digits. A hexadecimal number must be able to represent sixteen different states with only one digit. Since the decimal digits 0–9 can only represent ten states, the missing states are represented by letters. The letter *A represents* a 10 in hexadecimal, *B* represents an 11, and so on. Finally, the letter *F stands for* 15. Table 5.2 shows the mapping of decimal numbers to unique hexadecimal digits.

The conversion of a hexadecimal number into a decimal number follows the algorithm already known from Fig. 5.3 and is shown in Fig. 5.10 for hexadecimal numbers.

Fig. 5.10 Converting
hexadecimal numbers to
decimal numbers

$$\{9D\}_{16}$$
$$= \{9 \cdot 16^1\}_{10} + \{13 \cdot 16^0\}_{10}$$
$$= \{144\}_{10} + \{13\}_{10}$$
$$= \{157\}_{10}$$

$$42 : 16 = 2 \text{ Rest: } A \qquad \{42\}_{10} = \{ \ 0010 \ 1010 \ \}_2$$
$$21 : 16 = 0 \text{ Rest: } 2 \qquad \{42\}_{10} = \{ \quad 2 \quad A \quad \}_{16}$$

Fig. 5.11 Converting decimal or binary numbers to hexadecimal numbers

Conversely, decimal numbers can again be converted into a hexadecimal number by multiple division with base 16 and evaluation of the remainders, as already described for Fig. 5.4. However, there is a quick way to convert binary numbers into hexadecimal numbers. Since the base of the hexadecimal numbers equals 2^4, binary numbers can be divided into groups of four and then translated into hexadecimal numbers group by group. Always start with the bit that has the least significance. If the last group does not consist of four digits, it can be filled with 0 for positive numbers or with 1 for negative numbers. In simple terms, the last bit is always copied until the last group of four is full. Figure 5.11 shows this as an example for the conversion of the number 42.

Exercises

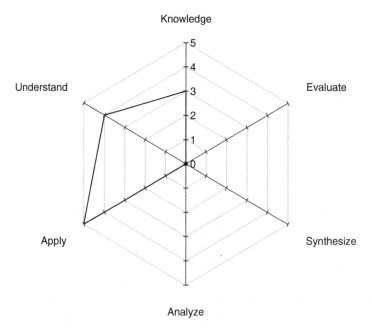

5.1 Variable Definition

What is the minimum information required to create a variable in a *C++* program?

5.2 Memory

How much memory is occupied by variables of type:

(a) *char*
(b) *short*
(c) *float*
(d) *int*
(e) *void*

5.3 Typecast

What does an *explicit typecast* do and what is the notation in a *C++* program?

5.4 Enumerations

Explain in your own words the advantages of enumerations!

5.5 Variable Definition in C++

Explain the differences between the terms *declaration*, *definition* and *initialization* in relation to variables in *C++*!

5.6   Number Systems

Justify why the number $\{10\}_B$ in any number system with base $B \geq 2$ is always exactly equal to the value of base B in the decimal number system!

5.7   The Duotrigesimal Number System

Make a table for all possible digits in the duotrigesimal number system with base 32 and assign the digits to their respective value in the decimal number system with base 10!

5.8   Print the Memory Requirement

Modify the *Hello World!* program to display the memory usage for different variable types on the screen using the *sizeof statement*!

The output should look like this:

```
Memory requirements of the variables:
bool = 1
char = 1
short = 2
int = 4
long = 4
long long = 8
float = 4
double = 8
long double = 8
```

5.9   Print the ASCII Codes

Print the numbers 97 through 105 to the screen and, as shown in Listing 5.2, convert the numbers to variables of type *char*! using an *explicit typecast*.

The output should look like this:

```
97 = a
98 = b
99 = c
100 = d
```

```
101 = e
102 = f
103 = g
104 = h
105 = i
```

5.10 Number System Conversion

Convert the following numbers into the respective target number system!

(a) $\{27\}_{10} = \{?\}_2$
(b) $\{11010010\}_2 = \{?\}_{16}$
(c) $\{6A\}_{16} = \{?\}_2$
(d) $\{127\}_8 = \{?\}_{10}$

5.11 Binary Addition and Subtraction

Convert the following numbers into the binary number system and perform the calculations in the binary number system! After the calculation, convert the results back into decimal numbers and check the results!

In this task, each task part counts as a single point for the overall score.

(a) $47 + 80 =$
(b) $4 - 73 =$

Branches

6

Short and Sweet

- Branching can be used to make decisions in programs.
- Decisions are made based on logical expressions.
- A logical expression knows only the two states *true or false*.
- The *if statement* allows you to distinguish between two states.
- The *switch-case statement* can distinguish between different constant states, which must be encoded with integers.

One of the most important tasks of a program is to make decisions. These can be simple decisions that only relate the value of one variable against another value, or complicated decisions that consist of several interconnected individual decisions. In this chapter, you will learn what operators exist to evaluate and combine expressions. You will also learn about the *if statement* and the *switch-case statement,* which can be used to implement alternative sequences in a program.

6.1 Operators for Comparisons and Logical Operations

In order to be able to make decisions using the *C++* programming language, statements, or expressions, must first be defined that can be either true or false. In *C++,* there is a separate variable type *bool* for this purpose, which can assume the values *true* and *false*. Since a *bool* variable occupies more than one bit of memory, the value 0 is interpreted as *false* and any other value as *true*. To formulate an expression using variables, comparison

B. Tolg, *Computer science to the Point*,
https://doi.org/10.1007/978-3-658-38443-2_6

Table 6.1 Comparison operators of the language

Operation	Description
A ==B	Checks whether two expressions A and B are identical
A > B	Checks whether expression A is larger than expression B
A < B	Checks if expression A is smaller than expression B
A >= B	Checks whether expression A is greater than or equal to expression B
A <= B	Checks whether expression A is less than or equal to expression B
A != B	Checks if expression A is not equal to expression B

Table 6.2 Logical operators of the language

Operation	Description
A && B	(logical AND) checks if both expressions are true
A ‖ B	(logical OR) checks whether at least one of the expressions is true
!A	(logical NOT) inverts the statement of the expression

operators are available for evaluation by the program. Table 6.1 shows a list of the comparison operators that can be used in C++.

Especially the first comparison operation, the ==, is a popular source of errors. In C++, the simple = assigns a value to a variable. The == checks whether two values are identical. However, because of the definition of *true* as not equal to 0 and *false* as 0, a value assignment can also be interpreted as a logical expression. If the variable is assigned the value 0, the expression would be *false,* otherwise *true*. Thus, there is no error message indicating that there is a high probability that what was intended was not evaluated.

Very often it happens that a single expression is not sufficient to describe a condition. In this case, several expressions must be linked together. In this case, the program must be informed by means of logical expressions in which way this linkage is to be carried out. Table 6.2 shows a list of the logical operators that can be used in C++.

There are also common sources of error in these logical operations. In mathematics, expressions like $5 < x < 10$ are used to express that x should be between 5 and 10. However, in the C++ language, this expression does not work. It would first check if $5 < x$ holds. This is either true or false, so the result would be *true* with a value of 1 or *false* with a value of 0. Then it would check if 0 or 1 is less than 10, which would be true in either case. The correct notation in C++ is $5 < x$ & & $x < 10$, when checking whether x *is* between 5 and 10. It is therefore important to always link individual expressions, as presented in Table 6.1, using the logical operators from Table 6.2 if more complex expressions are to be created.

Brackets should also be used to make the order of operations more visible. This can improve the maintainability and readability of the program code without much effort.

6.2   if-statements

In order to be able to decide between two alternatives based on an expression, the C++
language has the *if statement*. Figure 6.1 shows the representation of this form of branching in the form of a UML activity diagram. Immediately after the start, the first thing to do
is to get user input, which is to be stored in the variable *x*. The next step is the branching,
which takes place after the if statement. This is followed by the branch, which allows two
alternative program paths. In this example, the branching is to be carried out depending on
the value of *x* entered. If the value of *x* is less than or equal to 10, this is to be printed to
the console, otherwise it is to be printed that the value is greater than 10. Immediately
afterwards, the program is to terminate.

Figure 6.2 shows the syntax diagram for the *if statement*. The command always begins
with the keyword *if,* followed by a condition in round brackets. This is followed by either
a single statement, or a statement block, i.e. several statements enclosed by curly brackets.
Optionally, this can be followed by an *else,* which is also followed by a command or a
block. The syntax diagram for a block was presented in Fig. 2.4 on page 7.

Listing 6.1 shows the activity diagram shown in Fig. 6.1 as a C++ implementation. The
if branch exploits a comparison operation known from Sect. 6.1 to check whether the value
of *x* is less than or equal to 10. Since only a single statement follows, no block needs to be
inserted, i.e. the curly braces can be omitted. It is sufficient to formulate the output directly.
If the condition is not met, the output specified in the *else* branch is executed.

Fig. 6.1 Activity diagram for
a simple *if statement*

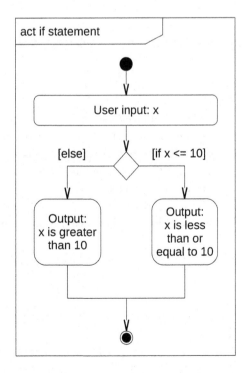

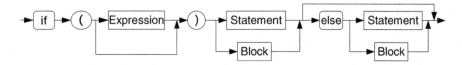

Fig. 6.2 Syntax diagram for the *if statement*

Listing 6.1 Implementing the Example Shown in Fig. 6.1 with an *if* Statement

```
1    #include <iostream >
2
3    using namespace std;
4
5    int main ()
6    {
7      // Variable definition and initialization
8      int x = 0;
9
10     // Reading a user input
11     // from the console into the variable x
12     cin > > x;
13
14     // Case distinction
15     if (x < = 10)
16        cout << "x is less than or equal to 10";
17     else
18        cout << "x is greater than 10";
19   }
```

Of course, with the *C++* language, multiple *if statements* can also be nested within each other. Listing 6.2 adds another distinction to the example. If the value of *x* is now less than or equal to 10, a further *if statement* distinguishes whether the value is really less than 10, in which case the corresponding output occurs, or whether the value is equal to 10, also with corresponding output.

Listing 6.2 Addition of Nested *if* Statements to the Example Shown in Fig. 6.1

```
1    #include <iostream >
2
3    using namespace std;
4
5    int main ()
6    {
7      // Variable definition and initialization
8      int x = 0;
```

```
9
10      // Reading a user input
11      // from the console into the variable x
12      cin > > x;
13
14      // Case distinction
15      if (x < = 10)
16        if (x < 10)
17          cout << "x is less than 10";
18        else
19          cout << "x is equal to 10";
20      else
21        cout << "x is greater than 10";
22    }
```

As can be seen in the example, it is not always easy to keep track of many *if statements* that are nested within each other. It is therefore advisable to consistently indent the areas that are within the *if statement*. The use of curly braces can also improve clarity, even if this is not absolutely necessary in this example.

Occasionally it happens that in programs a case distinction has to be made, which assigns an individual reaction to each case. Figure 6.3 shows such a case distinction for a very simplified travel agency (the possible destinations were chosen randomly and the

Fig. 6.3 Activity diagram for a case distinction

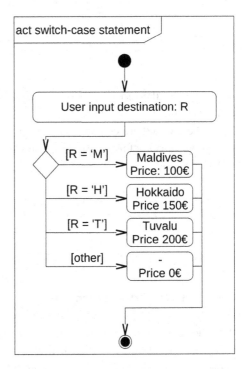

prices are, of course, nonsense). Suppose the user is asked to specify a destination, which directly determines the price of the trip. In this case, several *if-branches* can be placed in sequence to make this case distinction.

Listing 6.3 implements the example in Fig. 6.3 using several *if statements*. It should be noted that the *else if* statement is a normal *if* written in an *else* branch. So there are two independent statements. This could also be made clear by a different indentation.

Listing 6.3 Case Discrimination Using *if* Statements, Based on Fig. 6.3

```
1    int main ()
2    {
3      // Variable definition and initialization
4      double Price = 0.0;
5      char R = ";
6
7      // Reading a user input
8      // from the console into the variable R
9      cin > > R;
10
11     // multiple case distinction
12     if (R == 'M') Price = 100.0;
13     else if (R == 'H') Price = 150.0;
14     else if (R == 'T') Price = 200.0;
15     else price = 0.0;
16
17     return 0;
18   }
```

The type of the variable *R* is *char*, from which it follows that no numbers can be stored in this variable, but individual letters, or rather characters. If the content of the variable *R* is now to be checked in a comparison, it must be made clear that a variable is being compared with a character. For this reason, single quotation marks must be placed around the character.
Example:

```
if (R == a) // checks whether the value of the variable R
            // is identical to that of the variable a
if (R == 'a') // checks whether the value of the variable R
              // corresponds to the letter a
```

6.3   *switch-case statements*

With the help of the *switch-case statement,* case distinctions can be implemented particularly easily, as shown in Fig. 6.3. In this example, a very simple travel agency is to be implemented, which offers three randomly selected destinations. The prices have been

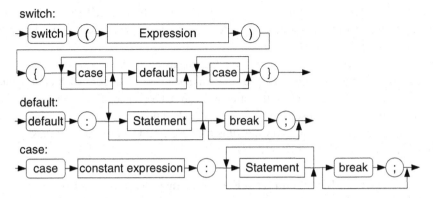

Fig. 6.4 Syntax diagram for the *switch-case statement*

intentionally chosen differently to allow for four different scenarios. As Listing 6.3 shows, case distinctions can also be made using the *if statement,* but this quickly becomes confusing if several conditions are to lead to the same result. The *switch-case statement* opens up many new possibilities here.

Figure 6.4 first shows the syntax diagram for the *switch-case statement*. The complexity of the diagram already suggests that there are many different ways to use the statement. First, the keyword *switch* must always be specified, followed by an expression in round brackets that can assume various values. It is important to note that these values must either correspond to an integer data type, or it must be possible to convert the expression into such a type. Of course, this condition also applies to variables of type *char,* or to the *enumerations* introduced in Sect. 5.3. A statement block is now defined in curly brackets, describing how to react to different cases.

Within the statement block, a new case distinction can be started with the statement *case*. This must be followed by a constant expression describing the value to be processed in this case. It must be underlined that only constant expressions are allowed, variables or even comparison operations must not be used. This must be followed by a colon. Now any number of statements can follow, which are relevant for this case. However, it is not necessary that statements actually follow. This has to do with the now following keyword *break,* which closes this case. The *break* ensures that the case distinction ends at this point. However, if it is omitted, the code intended for the next case distinction would be executed, and so on. This makes it possible to write several *case statements* in a row, all to be executed with the same code, or to add code successively depending on which case has been reached.

It may well be intentional that multiple instances of the *switch-case statement* result in the same behavior. Listing 6.5 on page 72 shows an example of this. This special case is called *fallthrough* in computer science. It should always be marked by a comment, since otherwise it is not possible to recognize whether the *fallthrough* is intended or whether a *break* was simply forgotten.

If none of the cases described by *case* occurs, the keyword *default* followed by a colon can be used to define a case that should occur in all other cases. This case can also be terminated by a *break*. Since the keyword *default* may only be specified once, but there may be any number of *case statements* before or after it, it was necessary to divide the syntax diagram into a total of three areas.

Listing 6.4 implements the example from Fig. 6.3 using a *switch-case statement*. Since different prices are required for each destination, a *break* was set after each case distinction. If an incorrect entry is made, the price is set to 0. Of course, this is not necessary in this simple program because the price was already initialized to 0 at the beginning. However, in a larger program, into which this example could be integrated, it makes sense to handle an error case in one way or another.

Listing 6.4 The Case Distinction from Fig. 6.3 Using a *switch-case statement*

```
1    int main ()
2    {
3       // Variable definition and initialization
4       double Price = 0.0;
5       char R = ";
6
7       // Entering the destination
8       cin > > R;
9
10      switch(R)
11      {
12        case 'M': // Trip to the Maldives
13          Price = 100.0;
14          break;
15        case 'H': // Trip to Hokkaido
16          Price = 150.0;
17          break;
18        case 'T': // Trip to Tuvalu
19          Price = 200.0;
20          break;
21        default: // Interception of the error case
22          Price = 0.0;
23          break;
24      }
25
26      // further program code
27
28      return 0;
29   }
```

Now, if all destinations cost the same price, the instructions and the *break instructions* could be omitted for the other cases. Listing 6.5 shows a modified program that distinguishes only between the case where any destination was specified or an incorrect entry was made.

Listing 6.5 Case Discrimination for Equal Prices Using *switch-case*

```
1   int main ()
2   {
3       // Variable definition and initialization
4       double Price = 0.0;
5       char R = ";
6
7       // Entering the destination
8       cin > > R;
9
10      switch(R)
11      {
12        case 'M': // Trip to the Maldives
13        case 'H': // Trip to Hokkaido
14        case 'T': // Trip to Tuvalu
15                  // Fallthrough
16          Price = 100.0;
17          break;
18        default: // Interception of the error case
19          Price = 0.0;
20          break;
21      }
22
23      return 0;
24  }
```

Exercises

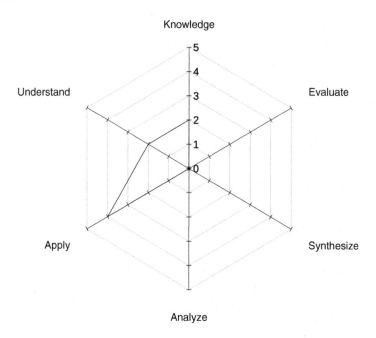

Network diagram for the self-assessment of this chapter

6.1   Comparisons

How can you check in *C*++ by logical comparison operations whether an expression *A*

(a) is equal to an expression *B*,
(b) is less than or equal to an expression *B*; or
(c) is not equal to an expression *B?*

6.2   Instruction Blocks

What is a statement block and how does it differ from a statement?

6.3   Comparisons

Explain the difference between = and == in the C++ language!

6.4   Branches

The C++s language has two different instructions that are used to make decisions in programs. What are the names of the two instructions and what are the differences.

6.5   *if statement*

Before you start programming this task, you should develop an activity diagram that describes the program flow!

You receive one point each for the activity diagram and the program.

Write a program that first prompts the user to enter his height k in meters. This value is to be stored in a variable of type *double* by a user input. Then the user is to enter his weight g in kilograms; this value is also to be stored in a variable of type *double*. Now the body mass index is to be calculated by the formula

$$double\ bmi = g\ /\left(k^*k\right);$$

The following outputs are now to be generated with the help of an *if statement:*

If the is *BMI* < 18.5 so the word *underweight* shall be printed.

Otherwise, it is to be checked whether the *BMI* < 25. In this case, the word *normal weight* is to be printed.

If neither of the first two cases applies, check whether the *BMI* < 30, then the word *overweight* should be printed.

In any other case, the word *obesity* should be issued.

6.6   *switch-case statement*

Before you start implementing this task, you should develop an activity diagram that describes the program flow!

You receive one point each for the activity diagram and the program.

Create an enumeration with the values *CELLAR, GROUNDFLOOR, LABS*, and *OFFICES* and name the enumeration *HOUSE*.

Now generate an output asking the user to select the floor he wants to go to. In doing so, print the various options, for example:

```
cout << "Cellar: " << CELLAR << endl;
```

Store the user input in a variable of type *int*. Then make a case distinction using a *switch-case* statement. The following outputs are to be produced:

Underground for the basement,
Ground level for the ground floor,
Over ground for laboratories and offices, as well as
Wrong entry! for every other case.

Loops

7

Short and Sweet

- Loops execute instructions repeatedly.
- The C++ language knows three types of loops.
- A distinction is made between head-controlled and foot-controlled loops:
  - head controlled
    while-loop
    for-loop
  - foot controlled
    do-while loop
- **Possible sources of error:**
  Especially with *while* and *do-while loops* there is the danger of an infinite loop, because often the incrementing of the loop variables is forgotten.

When implementing software, certain parts of the program often need to be executed repeatedly. Sometimes it is possible to specify exactly how often the code must be repeated. A simple example of this is when the values of a function are to be printed for all integers in the interval from 0 to 10. Figure 7.1 shows an activity diagram that illustrates the process. First, the value x is set to 0. Then the function value of x is printed. It does not matter for the example which function it is. Next, the termination condition is checked by comparing whether the value x is less than 10. If the condition is true, the program remains in the loop and again prints a function value for the now increased x. Otherwise, the program is terminated.

In other cases, however, only the termination condition is known, but not how long it takes for this condition to occur. For example, if the contents of an unknown file are to be

© The Author(s), under exclusive license to Springer Fachmedien Wiesbaden GmbH, part of Springer Nature 2023
B. Tolg, *Computer science to the Point*,
https://doi.org/10.1007/978-3-658-38443-2_7

Fig. 7.1 Activity diagram for
repeated output of func-
tion values

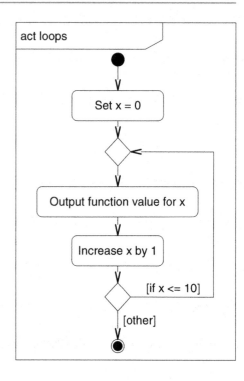

read in line by line, reading is to stop at the end of the file. Since the file is unknown, it
could just as easily contain a book as it could be empty.

In *C++*, various loop types are available that allow you to execute parts of the program
repeatedly. In principle, any loop type can be used for any application, but it can be very
inconvenient if the wrong loop type is chosen.

The given examples show that the termination condition plays an important role in
loops. This means that there is a state at the beginning of the loop, which is changed in the
course of the loop and which satisfies the termination condition at some point.

In *C++*, *a* distinction is made between head-controlled and foot-controlled loops. The
names are explained by the fact that a loop always begins with a loop head, followed by
the loop body and at the end the loop foot. The name always refers to the point in the loop
at which the termination condition is checked. *C++* knows three different types of loops,
which are introduced in the following.

7.1   *do-while loops*

The *do-while loop* is the only foot-controlled loop that exists in *C++*. Figure 7.2 shows the
syntax diagram for this type of loop. A *do-while loop* always begins with the keyword *do*,
followed by a statement, or statements within curly braces, to repeat. This is followed by

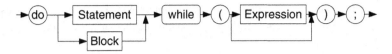

Fig. 7.2 Syntax diagram for the *do-while loop*

the *while statement* and, inside round brackets, an expression describing the termination condition. At the end of the *do-while loop*, there must always be a semicolon to complete the statement.

Foot-controlled loops have the property that the loop body is traversed first before the termination condition is checked. In contrast to the other loop types, the body of a *do-while loop* is traversed at least once, regardless of whether the termination condition is met or not.

Listing 7.1 converts the activity diagram shown in Fig. 7.1 into C++ code using a *do-while loop*.

Listing 7.1 Solving the Example Shown in Fig. 7.1 with a *do-while loop*

```
 1    #include <iostream >
 2
 3    using namespace std;
 4
 5    int main ()
 6    {
 7      // Variable definition and initialization
 8      int x = 0;
 9
10      // Loop head
11      do
12      {
13        // Loop body
14        cout << f(x);
15        x = x + 1;
16      }
17      // Loop foot
18      while (x <= 10);
19    }
```

Typical Applications

Since the body of the *do-while loop* is traversed at least once, it should be used in cases where this is exactly what is required. A typical example of this is user input.

If boundary conditions apply to a user input, such as that a value should be within a specified interval, then the query must be repeated until the input is within that interval.

How many repetitions the user needs to do this is not known. There must also be user input in each case, regardless of whether the current value of the variable would already satisfy the termination condition.

In Listing 7.2, a user input is expected in the interval of [0, 10]. The variable *x* is already initialized with 0 and would fulfill the termination condition. Nevertheless, at least one user input must occur before the termination condition can be checked.

Listing 7.2 User Annotation with Constraints with a *do-while loop*

```
1    #include <iostream >
2
3    using namespace std;
4
5    int main ()
6    {
7      // Variable definition and initialization
8      int x = 0;
9
10     // Loop head
11     do
12     {
13       // Loop body
14       cout << "Please enter a number "
15             << "between 0 and 10: ";
16       cin >> x;
17     }
18     // Loop foot
19     while (x < 0 || x > 10);
20   }
```

7.2   *while-loops*

The *while-loops* belong to the head-controlled loops. With this loop type, the termination condition is checked before the loop body is executed. It is therefore possible that the loop body is never reached if the termination condition has already been fulfilled at the beginning. Figure 7.3 shows the syntax diagram for this loop type.

Fig. 7.3 Syntax diagram for the *while-loop*

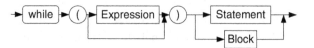

The loop is introduced with the keyword *while,* followed by an expression in round brackets that describes the termination condition. This is followed by the loop body, which consists of either a single statement, or a block of statements within curly braces. Listing 7.3 shows the implementation of the activity diagram shown in Fig. 7.1 using a *while-loop.*

Listing 7.3 Solving the Example Shown in Fig. 7.1 with a *while loop*

```
1    #include <iostream >
2
3    using namespace std;
4
5    int main ()
6    {
7      // Variable definition and initialization
8      int x = 0;
9
10     // Loop head
11     while (x <= 10)
12     {
13       // Loop body
14       cout << f(x);
15       x++;
16     }
17   }
```

Typical Applications

The *while-loops* are commonly used in situations that meet two conditions. First, there is a clearly defined termination condition that establishes the end of the loop. Second, it is unknown how many passes of the loop are needed to reach this end.

An illustrative example is reading data from a file. The termination condition is clearly defined: When the end of the file has been reached, reading must stop. However, the number of passes is not known unless there is additional information about the file. The file could be empty or contain several gigabytes of data.

Another, less demanding, example comes from the field of approximation methods in numerical mathematics. Approximation methods are always used when no exact solution can be found, or it would be extremely difficult to find one.

Listing 7.4 shows a very simple approximation procedure for finding the root of the number 8. Without thinking further, one can assume that the solution will be found somewhere within the interval [0, 8]. Let the lower bound be called *a,* and the upper bound *b.* The program should now push the limits towards each other in such a way that they approach the correct result from above and below.

Listing 7.4 Example of a Simple Approximation Procedure using a *while loop*

```
1    #include <iostream >
2
3    using namespace std;
4
5    int main ()
6    {
7      // Variable definition and initialization
8      double a = 0.0;
9      double b = 8.0;
10     double c = 0.0;
11
12     // Loop head
13     // The termination condition checks the distance
14     // of the two interval limits
15     while (b - a > 0.000001)
16     {
17       // Loop body
18
19       // Calculation of the value in the
20       // Middle of the interval
21       c = (a + b) / 2.0;
22
23       if (c*c > 8.0)
24         // Replacing the upper limit
25         b = c;
26       else
27         // Replacing the lower limit
28         a = c;
29     }
30
31     cout << c << endl;
32   }
```

The termination condition is then very simple: The program should terminate if the distance from the upper to the lower limit is smaller than a specified maximum error. In this example, the maximum error is 0.000001.

However, the number of loop passes required to achieve this goal is not known precisely without further consideration. If the boundaries of the interval are already close enough to each other, then no single pass of the loop would be necessary. If the boundaries are unfavorable, a very large number of passes may be necessary.

In the loop body, the interval is reduced with each pass. For this purpose, the middle of the current interval is calculated and stored in the variable c The value in c is thus the test value for the searched root of the number 8. If c^2 is now greater than the value 8, then it follows that c is greater than the searched root. In this case, the upper interval boundary b

is moved to *c*. Otherwise, the lower interval boundary *a* is shifted to *c*. With each further loop pass, one of the two limits converges to the searched result until the termination condition is finally fulfilled.

7.3   *for-loops*

The for-loop also belongs to the head-controlled loops. The syntax diagram is shown in Fig. 7.4. It is easy to see that its structure differs from that of the other two loop types. The keyword *for* is followed by three statements, each separated by semicolons. The individual statements, labeled *A*, *B,* and *C* in the figure, are optional. However, the semicolons must be specified in each case.

- Instruction *A* is executed before the loop passes begin. Very often a count variable is initialized here.
- Instruction *B* contains the termination condition, which is checked before each loop pass.
- Instruction *C* is executed after each loop pass. In many cases, the variable initialized in instruction *A* is changed here, so that the termination condition is fulfilled at some point.

Since the *for-loop* is very often used to count up or down variables, it is also called a counting loop. Listing 7.5 shows the implementation of the activity diagram shown in Fig. 7.1 using a *for-loop*.

Listing 7.5 Solution of the Example Shown in Fig. 7.1 with a *for loop*

```
1     #include <iostream >
2
3     using namespace std;
4
5     int main ()
6     {
7        // Loop head
8        for (int x = 0;x <= 10;x++)
9        {
10          // Loop body
11          cout << f(x);
12       }
13    }
```

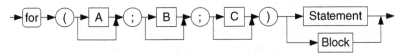

Fig. 7.4 Syntax diagram for *the for-loop*

The first instruction defines and initializes the loop variable x with the value 0. As a termination condition, the second instruction specifies that the loop should run as long as the value of x is less than or equal to the value 10. The third statement x++ is a very abbreviated notation of the statement $x = x + 1$. If the value of x is to be incremented with another value, e.g. 2, the statement $x + = 2$ could also be used as an abbreviated notation.

Typical Applications
The representation of mathematical formulas, as shown in Listing 7.5, is already a very typical example of the use of the *for-loop*. Even though the same task can be performed with any type of loop, it is noticeable that the implementation with the *for-loop* turns out to be very concise and clear.

Other typical use cases for *for-loops* are described in Sect. 8 about the so-called arrays. Many operations performed on arrays require the use of loops, and the *for-loop* is often the best choice.

Multidimensional formulas can also be represented well by *for-loops*. For this purpose, several loops can be nested within each other. This works with all loop types presented and will be carried out here with the *for-loop* as an example. Listing 7.6 shows such an implementation.

Listing 7.6 Representing a Function with Two Variables Using *for loops*

```
1    #include <iostream >
2
3    using namespace std;
4
5    int main ()
6    {
7      // Loop for the y-values
8      for (int y = 0;y < 5;y++)
9      {
10       // Loop for the x-values
11       for (int x = 0;x < 5;x++)
12       {
13         // Output of the function values
14         cout << "(" << x << ", " << y << ")" << "\t";
15       }
16       cout << endl;
17     }
18   }
```

The outer loop traverses the rows of the plot and increments the variable y on each pass. Inside the outer loop two things happen: The inner loop runs through each column and

a	b	c
(0, 0) (1, 0) (2, 0) (3, 0) (4, 0)	(0, 0) (1, 0)	(0, 0) (1, 0) (3, 0) (4, 0)
(0, 1) (1, 1) (2, 1) (3, 1) (4, 1)	(0, 1) (1, 1)	(0, 1) (1, 1) (3, 1) (4, 1)
(0, 2) (1, 2) (2, 2) (3, 2) (4, 2)	(0, 2) (1, 2)	(0, 2) (1, 2) (3, 2) (4, 2)
(0, 3) (1, 3) (2, 3) (3, 3) (4, 3)	(0, 3) (1, 3)	(0, 3) (1, 3) (3, 3) (4, 3)
(0, 4) (1, 4) (2, 4) (3, 4) (4, 4)	(0, 4) (1, 4)	(0, 4) (1, 4) (3, 4) (4, 4)

Fig. 7.5 Output of the various programs for function output. (**a**) Listing 7.6, (**b**) Listing 7.7, (**c**) Listing 7.8

prints the value of the x- and the *y-coordinate* in brackets. To keep a little space between the outputs, \t adds another tab.[1]

After each run of the inner loop, an end of line is brought about in the outer loop. Figure 7.5a shows the result.

7.4   *continue* and *break*

When running through loops, it can sometimes make sense to interrupt or skip the run. For example, if an error is detected while a loop is being run, it no longer makes sense to run the rest of the loop. An error message should be displayed and the loop should be interrupted. In principle, this could also be realized by an *if-query*, which would execute a part of the loop content only if the error case has not occurred. However, this can sometimes be cumbersome to implement.

The *break* command can be used to interrupt a loop immediately. If several loops are nested within each other, only the loop in which the *break statement* is located is affected. Example Listing 7.7 modifies Listing 7.6 so that the inner loop is broken when the third column is reached. Figure 7.5b shows the modified output of the program.

Listing 7.7 Modifying Listing 7.5 Using *break*

```
1    #include <iostream >
2
3    using namespace std;
4
5    int main ()
6    {
7      // Loop for passing through the lines
```

---

[1]The text console is divided into several columns, which can be jumped to by the tabulator. This gives the output a tidier appearance. However, if the output of the function is of different lengths, it can happen that one of the outputs extends beyond a tabulator position, so that an offset occurs for the subsequent outputs. In such cases, the length of the output should be limited.

```
8      for (int y = 0;y < 5;y++)
9      {
10       // Loop for passing through the columns
11       for (int x = 0;x < 5;x++)
12       {
13         // Interruption of the loop in the
14         // third column using break
15         if (x == 2) break;
16
17         // Output of the function values
18         cout << "(" << x << ", " << y << ")" << "\t";
19       }
20       cout << endl;
21     }
22   }
```

The *continue command* behaves similarly to the *break command*, but only skips the current pass of the loop and allows the loop itself to continue. All commands that follow the continue command are skipped.

Listing 7.8 modifies Listing 7.6 to skip the output of the third column. Figure 7.5c shows the correspondingly modified output.

Listing 7.8 Modifying Listing 7.5 Using *continue*

```
1    #include <iostream >
2
3    using namespace std;
4
5    int main ()
6    {
7      // Loop for passing through the lines
8      for (int y = 0;y < 5;y++)
9      {
10       // Loop for passing through the columns
11       for (int x = 0;x < 5;x++)
12       {
13         // Skip the loop contents of the
14         // third column using continue
15         if (x == 2) continue;
16
17         // Output of the function values
18         cout << "(" << x << ", " << y << ")" << "\t";
19       }
20       cout << endl;
21     }
22   }
```

Exercises

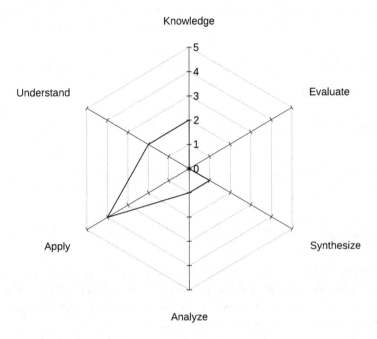

Network diagram for the self-assessment of this chapter

7.1   Loops

Name the three ways to implement loops in *C++* and classify them as head or foot controlled loops.

7.2   Use Cases

Name a typical use case for each of the three loops.

7.3   Loop Types

Explain in your own words the difference between head-controlled and foot-controlled loops!

7.4   Endless Loops

Explain how an infinite loop is created!

7.5   Elevator

Take the program from Exercise 6.6 and modify it so that you are asked for the floor to which you want to go until you make an invalid entry. The program should also remember the floor you are currently on. If you select the current floor again, the message *"You are currently here"* should appear instead of the usual text.

7.6   Printing of the Character Mapping Table

In this task, each task part counts as a single point for the overall score.

(a) Write a program that loops through the numbers 0 to 255 and outputs them to the console. In addition, as in Exercise 5.9, the values are to be converted to *char variables* by a *typecast* and also output.
(b) Complete your program so that the output is in 2 columns.
(c) Complete your program so that the output is in 3 columns.

7.7   Program Analysis

Look at the following program and try to figure out what this program does. The point is not to describe the program line by line, but to get an accurate picture of what the goal of the program shown is. Of course, you can just type out the program and try it out, but that is not the goal of the exercise. The program uses instructions that you do not yet know. You may, of course, research them.

Listing 7.9 Program with Unknown Task

```
1    #include <iostream >
2
3    using namespace std.;
4
5    int main ()
6    {
7      const int N = 21;
8
9      for (int y = 0; y < N; y++)
10     {
11       for (int x = 0; x < N; x++)
12       {
13         int dx = x - N / 2;
14         int dy = y - N / 2;
15
16         if (sqrt(dx * dx + dy * dy) < N*0.4 &&
17             sqrt(dx * dx + dy * dy) > N*0.1)
18         {
19           cout << "*";
20         }
21         else
22         {
23           cout << "";
24         }
25       }
26       cout << endl;
27     }
28
29     return 0;
30   }
```

7.8   Output of the Character Mapping Table

Add a user input of the number of columns s to your program from Exercise 7.5. The value should be an integer and lie in the interval $[1; 10]$. Then the output is to be in s columns.

Arrays

8

Short and Sweet

- Arrays store N data under a common name.
- The entries of an array are called elements.
- An array can extend into several dimensions.
- Also texts are nothing else than an array of N characters.
- Indices are used to obtain individual data.
- Loops are often[1] used for access.
- The smallest index of a field is always 0.
- The highest index of a field is always $N - 1$.
- **Possible sources of error:**

No check is made to see if the index is within bounds when the array is accessed.

If programs are to process data, it is often necessary to manage many values that all have the same variable type. If, for example, a temperature curve is to be recorded in an experiment over a period of one minute, with a new temperature value being measured every 5 s, this results in a table of values as shown in Table 8.1.

Now this problem is still very manageable. It would be possible to create 13 variables of the type *unsigned int* for the time and of the type *double* for the temperature. But this

---

[1] The *for loop is* used very often for several reasons. On the one hand, a fixed number of elements must always be passed through for arrays. Secondly, the *for loop* with its loop header offers a very compact representation of all relevant information. In addition, the loop variable is changed in the loop header, thus reducing the risk of it being forgotten.

© The Author(s), under exclusive license to Springer Fachmedien Wiesbaden GmbH, part of Springer Nature 2023
B. Tolg, *Computer science to the Point*,
https://doi.org/10.1007/978-3-658-38443-2_8

Table 8.1 Measured temperature curve for an experiment

Time (s)	Temperature (°C)
0	1.00
5	1.42
10	2.01
15	2.86
20	4.06
25	5.75
30	8.17
35	11.59
40	16.44
45	23.34
50	33.12
55	46.99
60	66.69

approach would not be very flexible and at the latest if the measurement should cover several days, the resulting program would no longer make sense with this approach.

In mathematics, there is a very simple solution to this problem. The name of a series of measurements is given an index to make it clear that there are many values involved. The name T_i, with $i = 1, ..., N$ could thus stand for the measured temperatures and T_4 for the fourth measured temperature. In our example $N = 13$, would thus be the number of measured values.

The same approach was chosen for many programming languages. In computer science, the term for a group of values that are all addressed with a name and an index is *array*,[2] the values within the array are called elements. For the programming language, it is important that the index can be clearly distinguished from the name of the array. In the C++ programming language, the index is therefore always written in square brackets after the name of the array.

The T_i known from mathematics was written in C++ thus $T[j]$, but with the difference that in C++ the index j always starts at 0 and ends at $N - 1$. The fourth value of the measurement T_4 would therefore be $T[3]$ in C++.

Listing 8.1 creates various arrays and shows different ways to initialize them.

Listing 8.1 Defining and Initializing Arrays

```
1    #include <iostream >
2
3    using namespace std;
```

---

[2] In fact, both terms are used absolutely equally. In my personal use of language, however, I use the word *array* much more frequently. For the book, however, I didn't want to keep switching between German and English terms, so I'll stick with the term field. However, it is best to get used to both terms.

```
4
5    int main ()
6    {
7      // define a constant for the array size
8      const unsigned int N = 13;
9
10     // Initialization of an array with predefined
11     // Values
12     unsigned int time[N] = { 0, 5, 10, 15
13                            , 20, 25, 30, 35
14                            , 40, 45, 50, 55, 60};
15
16     // Initialization of an array with the value 0
17     double T[N] = { 0.0 };
18
19     // Initialization of the array with a
20     // for -loop
21     for (int i = 0 ; i < N ; i++)
22     {
23       cout << "Please enter the temperature after "
24            << time[i] << "s on: ";
25       cin >> T[i];
26     }
27   }
```

In line 8, an auxiliary constant N is first initialized with the value 13 for the array size. This is very useful if the array is to be looped through later. In this case, the auxiliary constant N can always be used as a reference. If the array size is to be adjusted, an adjustment at one point is sufficient. Variable values are not allowed when defining arrays, so the variable must be provided with the keyword *const*. Of course, instead of the constant N, the value 13 can always be written. This makes it clear that the length of such an array cannot easily be changed at runtime. How arrays with variable size are created is explained in Chap. 11.

For the example with the measured temperatures, we need an array in which we can store the times for the measurements. To define an array, we need to write the number of elements that the array should consist of in square brackets after the names of the variables, as in line 11.

At this point there are often misunderstandings. According to this definition, the array *time* consists of N (i.e. 13) elements. However, the valid indices are the values 0 to 12.

Since the values were specified in this example, the array can also be initialized immediately during the definition in line 12. One possibility of initialization is to assign the values to the array in curly brackets and separated by commas. However, this notation is only allowed for initialization. If only some values are to be initialized, fewer values can be placed in the parentheses. The remaining values are then initialized to 0. However, it is never permitted to write too many values in the brackets.

In line 17, the array for the measured temperatures is also defined with 13 elements. Since the temperatures are not yet known, all elements of the array are initialized with 0 by the empty curly brackets.

From line 21, the values of the array T are read in using a *for loop*. In the loop header, the count variable i is defined and initialized so that it runs through the value range from 0 to $N - 1$. Within the loop, an explanatory text is printed and the value for the number of seconds is taken from the array *time* at the *ith* position. Then, in line 25, the user input is stored in the *ith* element of the array T.

After the program has been run, the values stored in T can be used for various further calculations. For example, it would be possible to determine the mean or standard deviation of the temperature.

The values of an array are always arranged directly one after the other in memory. If an array consists of 10 elements of type *int* and an *int* occupies 4 bytes of memory, the array will occupy $10 \cdot 4$ bytes $= 40$ bytes of memory. The exact amount of memory used can be determined using the *sizeof* function. For example, *sizeof(T)* in Listing 8.1 returns the value $13 \cdot 4$ bytes $= 52$ bytes.

8.1   Strings

In Chap. 5 the variable type *char* was already introduced, which can store single letters. Of course, for many applications it is not sufficient to store only single letters, instead whole words are needed.

Now a word is nothing else than an ordered set of letters, or more generally, characters. In principle, this is exactly what is created by an array. In the C programming language, words are also represented in exactly this way, by a simple array with the variable type *char*. These arrays are called *C-strings*. In the C++ programming language, there is a separate *string* data type for this purpose, which provides additional functionality. However, *C-strings* are still used, so the *string* variable type is not a replacement, but more an addition that simplifies the use of strings.

Since a *C-string* is a normal array, its length cannot be changed easily during the runtime of the program. Therefore, the length of such a *C-string* is always defined in such a way that in the special application case every possible text would fit into it in any case. Now the array is too large for many texts and it must still be defined when the text ends within a too large array. This is achieved by writing the value 0 after the text in the array. This is done automatically in many cases, such as value assignment. Since the 0 marks the end of a text, we also speak of null-terminated *strings*.

Also the length of a text cannot be derived immediately from the size of the array, because the array size always represents only the maximum size. To find out the length of a text, a loop would have to go through the text and count all characters until the value 0 is found. Listing 8.2 shows how the length of a text in a *C-string* can be determined.

Listing 8.2 Using *C strings*

```
1    #include <iostream >
2
3    using namespace std;
4
5    int main ()
6    {
7      // Auxiliary variable , to determine the length
8      int length = -1;
9
10     // Assignment of a sample text to a C - String
11     char text [1024] = "This is a sample text ";
12
13     for (int i = 0 ; i < 1024 ; i++)
14     {
15       // The first 0 in the array is used as the length of the text.
16       // saved
17       if (text[i] == 0)
18       {
19           length = i;
20           break;
21       }
22     }
23
24     cout << "The text: " << text << endl << "consists of "
25          << length << " characters !" << endl;
26   }
```

In line 8, an auxiliary variable of type *int* is first defined and initialized with the value -1. In line 11, an array with 1024 elements of type *char* is then defined and initialized with a sample text.

From line 13 onwards, a *for loop* follows, which runs through the array once completely. If the value of the *text* array at position *i* is 0, *length* is set to the current value of *i* and the loop is terminated with the statement *break;*. This ensures that only the first 0 in the text array causes the value of the *length* variable to change. It should also be noted that *text[i]* is compared with the number 0 as the end for the text. Should be compared to the text character '0', the number would have to be enclosed in single quotes to mark it as *char.*

The same task is now solved in Listing 8.3 using a *C++ string.*

Listing 8.3 Using *C++* strings

```
1    #include <iostream >
2    #include <string >
3
4    using namespace std;
5
6    int main ()
7    {
8      // Assignment of a sample text to a C - String
9      string text = "This is a sample text ";
10
11     cout << "The text: " << text << endl << "consists of "
12          << text.length () << " characters !" << endl;
13   }
```

In order to be able to use the data type *string*, the library *string* must first be included, as in line 2. Like many other standard *C++* libraries, *string* also uses the *namespace std*. It is therefore advisable to *use namespace std;* to ensure that the *namespace* does not have to be named with every call. In line 9, a variable of type *string* is now created and initialized with the same text as in the example Listing 8.2.

In line 11, the result is printed directly, since the variable type *string* provides a function *length()*, which determines the length of the contained text. The call is made in line 12 by *text.length()*. Functions are described in more detail in Chap. 9.

Although the variable *text* in Listing 8.3 is now a *string*, the individual letters of the text can still be accessed using the square brackets. The *for loop* from Listing 8.2 would therefore also work in Listing 8.3.

When processing character strings and especially when sorting, the fact that each character is assigned a number that can be used synonymously is helpful. Table 8.2 shows an excerpt from the *ASCII table*.

As can be seen from the table, the upper case letters are in the interval [65; 90] and the lower case letters are in the interval [97; 122]. The characters '0' to '9' are in the interval [48; 57].

Since each letter and number sign is now assigned to a number, the computer can work internally with these numbers. So the query *if (A' < B') ...* would actually be true, since it corresponds to the query *if (65 < 66) ... .*

8.2   Multidimensional Arrays

For various tasks, it may be necessary for an array to have more than one dimension. For example, if the data of a computertomograph are to be analyzed, three dimensions are needed to assign a value for the absorption coefficient to each point in space. In

Table 8.2 Extract from the ASCII character assignment table

Characters	Code	Characters	Code	Characters	Code
'0'	48	'A'	65	'a'	97
'1'	49	'B'	66	'b'	98
'2'	50	'C'	67	'c'	99
'3'	51	'D'	68	'd'	100
'4'	52	'E'	69	'e'	101
'5'	53	'F'	70	'f'	102
'6'	54	'G'	71	'g'	103
'7'	55	'H'	72	'h'	104
'8'	56	'I'	73	'i'	105
'9'	57	'J'	74	'j'	106
...	...	...	...	...	...
...	...	'X'	88	'x'	120
...	...	'Y'	89	'y'	121
...	...	'Z'	90	'z'	122

mathematics, there are functions that assign a value for the *z-coordinate* to each *x-coordinate* and to each *y-coordinate,* and so on.

To create an array with multiple dimensions in *C++*, simply specify all dimensions one after the other in square brackets when defining a new array. The definition *int array2d[5] [10];* would thus create a two-dimensional array with 5 · 10 elements. Accessing an array with two or more dimensions is done analogously to one-dimensional arrays by specifying the respective indices in square brackets. Here, too, the counting of the indices starts at 0.

As an example, the mathematical function $z = \sqrt{x^2 + y^2}$ is to be stored in a two-dimensional array. Listing 8.4 shows an example implementation.

Listing 8.4 Applying a multidimensional array in *C++*

```
1    #include <iostream >
2    // in the cmath library there are
3    // mathematical functions , e.g. sqrt
4    // to calculate the root
5    #include <cmath >
6
7    using namespace std;
8
9    int main ()
10   {
11      // Auxiliary variables for the expansion of the array
12      const int Y = 50;
13      const int X = 50;
14
15      // Definition of the two-dimensional array
```

```
16        double z[Y][X] = {0.0};
17
18        // Calculation of the function values for all (x, y)
19        // Coordinate pairs
20        for (int y = 0; y < Y; y++)
21        {
22          for (int x = 0; x < X; x++)
23          {
24            z[y][x] = sqrt(x*x + y*y);
25          }
26        }
27  }
```

The two-dimensional array required for the results of the calculation is defined in line 12. To define the size of the two dimensions, the auxiliary variables X and Y are used, which were initialized in lines 8 and 9.

From line 16, the calculation of the function values begins with the help of two nested loops. In line 20, the function values are assigned to the individual array elements; access is provided by specifying the indices in square brackets in each case.

8.3   Multidimensional Arrays with Strings

In programs that are to process texts, it can come to the situation that inputs are distributed over several variables of the type *string*. For example, it may be necessary to manage a series of inputs that must then be sorted according to certain criteria. In these cases, the program also works in principle with multidimensional arrays. If *C-strings* are used, it is obvious that a multidimensional array is involved. The line *char words[N][M]* would produce an array capable of holding *N* words of length *M*. The example Listing 8.5 shows how multidimensional strings can be used.

Listing 8.5 Multidimensional strings with *C strings*

```
1    #include <iostream >
2    //# include <string >
3    using namespace std;
4
5    int main ()
6    {
7      // Auxiliary variables for the expansion of the array
8      const unsigned int N = 5;
```

```
9      const unsigned int M = 255;
10     // Initialization of the array
11     char words[N][M]
12        = { "this", "is", "a", "test", "!" };
13     // string words[N]
14     // = { "this", "is", "a", "test", "!" };
15
16     // Output of all words of the array
17     for (int i = 0; i < N; i++)
18     {
19       cout << words[i] << endl;
20     }
21
22     // Output of all words of the array Variant 2
23     for (int i = 0; i < N; i++)
24     {
25       int j = 0;
26       while (j < N && words[i][j] != 0)
27       {
28         cout << words[i][j];
29         j++;
30       }
31       cout << endl;
32     }
33   }
```

In line 11, a multidimensional array of type *char* is created and initialized with a series of words. The words are separated with commas and written in double quotes so that they are recognized as *strings*. The array has the size 5 · 255, as defined by the constants in lines 8 and 9. Alternatively, C++ *strings* can be used in this example. To do this, line 11 must be commented out and line 13 activated and the *preprocessor directive #include < string >* added.

If the words are now simply to be printed, there is a simplified notation to do this. Since the *C-strings* always end when the value 0 is in the array, a simple loop as in line 17 is sufficient to realize the output, since the end of the word can be recognized by the printing operation.

In line 23, the output of the complete words also takes place. The result is identical to that achieved in line 17 by a single loop. In this case, however, each letter is printed individually. The inner *while loop* runs through the respective words until either the end of the dimension is reached or the value 0 is found in an element.

Since the type *string* offers many additional functions compared to *C-strings* and is easier to use, it is advisable to use this data type, especially at the beginning.

Exercises

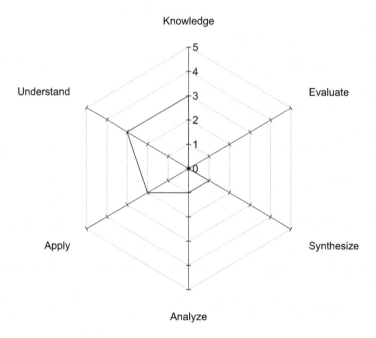

Network diagram for the self-assessment of this chapter

8.1   Indices

In which interval are the indices of an array defined by the line *int array[15];* located?

8.2   Strings

By which character is the end of a text in a *C-string* marked and what are these *strings* therefore called?

8.3   ASCII Table

What is represented in the ASCII table?

8.4   Letter Comparison

Explain why a comparison of two letters, as in *if('a' < 'b')* … works!

8.5   Function Values

In a program, the values of a function are to be calculated and printed. After this output, the function values are no longer required. Is an array required for this program?
     Give reasons for your answer!

8.6   Numbers and Characters

Explain the difference between the number 9 and the character '9' in

8.7   Random Numbers

Create a program in which an array of type *int* with $N = 100$ elements is created. Now *loop* through each element of the array with a *for loop* and assign the value *rand() % 1000* to each element. The *rand()* function generates a random integer. The modulo operation % calculates the remainder when divided by the number 1000. The result is a number that lies in the interval [0; 999].
     Add up the numbers in a second loop and divide the result by *N*. Enter the result of your program with the hint

Mean value of random numbers: x

on the console. Replace the *x* with the calculated value.

8.8   Largest Initial Letter

Write a program that reads 10 words into an array of the data type *string*. Afterwards, the words are to be searched within your program, so that the word with the first letter is found and output, which can be found last in the alphabet.

8.9   Program Analysis

Analyze the following program. To do this, try to find out what the individual program lines do in terms of content and deduce the task of the program.

Commands have been used that you do not know yet. Try to research them.

Do not type the program, but try to understand what is happening without assistance!

```
1    #include <iostream >
2    #include <time.h>
3
4    using namespace std;
5
6    int main ()
7    {
8      srand(time (0));
9
10     const int N = 1000;
11     double values[N] = { 0.0 };
12
13     for (int i = 0; i < N; i++)
14     {
15       values[i] = (( double )(margin () % 1000)) / 100.0;
16     }
17
18     for (int i = 0; i < N; i++)
19     {
20       for (int j = 0; j < N - 1; j++)
21       {
22         if (values[j] > values[j + 1])
23         {
```

```
24              double h = values[j];
25              values[j] = values[j + 1];
26              values[j + 1] = h;
27           }
28        }
29     }
30
31     for (int i = 0; i < N; i++)
32     {
33        cout << values[i] << endl;
34     }
35
36     return 0;
37  }
```

8.10   Word Lengths

Write a program that generates $N = 1000$ words with a random length of at least 3 and at most 10 lowercase letters and stores them in an array.

Then your program should determine the mean word length and the standard deviation of all word lengths and print them together with the words on the screen.

The formula for the arithmetic mean is:

$$\bar{x} = \frac{1}{N} \sum_{i=0}^{N-1} x_i \tag{8.1}$$

The formula for the standard deviation is:

$$s = \sqrt{\frac{1}{N-1} \sum_{i=0}^{N-1} (x_i - \bar{x})^2} \tag{8.2}$$

Functions

9

Short and Sweet

- A function is declared by:
  - Return type
  - Name
  - Parameters in round brackets.
- Functions can be used to structure programs.
- Functions can be used again and again.

The more complex the tasks that are to be solved by computers become, the more complex the programs that are to solve these tasks also become.[1] In complex software systems, there are many tasks that must be executed at different points in a program. For example, it may be necessary to print an error message at several places, which should always follow a certain pattern. For example:

```
ERROR: A file could not be opened!
Do you want to continue (y/n)?
```

The beginning of the line should always be the same. However, the error message at the end may change, depending on the specific case.

Now, of course, it is easy to copy these few lines into the program wherever an error message is to be printed. After that, only the text of the error would have to be changed.

---

[1] Malicious tongues claim that many of the problems solved by computers would not exist without them.

But what happens if the structure of the entire output is to change? From now on, the out-put should always look like this, for example:

```
ERROR:
-------
A file could not be opened!
-------
Do you want to continue (y/n)?
```

In this case, you search your entire program for the error outputs in order to change them. Of course, depending on how complex your program already is, there's always a risk that you'll miss some output or other. And there are much more complex parts of a pro-gram that can be repeated.

It is obvious that repetitive program sections should be written only once so that they can be reused. In addition, these related sections should be assigned names so that it is easy to see what they do. And as the example already shows, it may be necessary for these program sections to be passed values, or to return values. All these tasks are performed by functions.

Figure 9.1 shows the syntax diagram for functions in C++.

The function header always consists of three parts:

- The return type determines what kind of value is returned by the function. The function always has only one return type and can only return one value of this type. Basically, a function here behaves like a corresponding variable but with the difference that the returned value is still calculated by the function. There may also be functions where it makes no sense for them to return a value. In that case, a return value must still be specified, which is then *void*. The word *void* stands as a placeholder if no data type is to be returned.
- The name of the function determines how the function will be called in the future within the program. It makes sense to choose a name that describes the function's task as briefly as possible.
- Within round brackets, parameters can be specified to be passed to the function. It may happen that this is not necessary for certain functions. In this case, however, the paren-theses must still follow the name of the function. If several parameters are to be passed, they must be separated by commas.

Fig. 9.1 Syntax diagram for functions

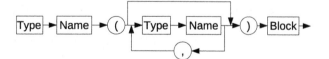

Listing 9.1 shows the implementation of a function that implements the error message from the previous example.

Listing 9.1 An Error Message Function

```
1    #include <iostream >
2    #include <string >
3
4    using namespace std;
5
6    // Implementation of the error function
7    bool error(string errormessage)
8    {
9      char result;
10
11     cout << "ERROR :" << endl;
12     cout << "-------" << endl;
13     cout << errormessage << endl;
14     cout << "-------" << endl;
15     cout << "Do you want to continue (y/n)?" << endl;
16     cin >> result;
17
18     if (result == 'y')
19     {
20       return true;
21     }
22
23     return false;
24   }
25
26   // Main function
27   int main ()
28   {
29     // ...
30     bool goOn = error (" A file could not ...!");
31
32     if (goOn == false)
33     {
34       return 10;
35     }
36     // ...
37   }
```

If a function is to be written, it must always be clarified first which value the function is to return. In the example, the error message is to query whether the program should continue after the error. Of course, it would be possible to pass directly through the result of the user input, but if the required input changes at some point, the subsequent processing may no longer be correct. For this reason, it makes sense to choose a slightly more abstract return. The variable type *bool* lends itself to this, as it only distinguishes between *true* and *false*. Within the function it can then be decided what is true or false in this case.

After that, a meaningful name for the function must be found. Since the example is about an error output, the name *error* is suitable. It is important that the function is declared[2] before it is used for the first time, in this case before the *main function*.

Now it must be decided which parameters the function requires. In this example, the function fulfills a specific task, it is to print an error message. This error output should always follow the same pattern, but the text of the error message depends on which error has occurred. This is the decisive criterion for a parameter. The function itself cannot know which error has occurred, consequently it cannot decide on the text of the error message. This information must come from outside, and that is a function parameter.

The function *error* was implemented after all these considerations. In line 9 an auxiliary variable is created within the function to read in the later user input. A common question is whether it is at all possible to create a variable of type *char* in a function of type *bool*. And the answer is that one has nothing to do with the other. The type of the return value only refers to the value that the function will eventually return to the place from where it was called. The command for this is *return*. Lines 20 and 23 specify the return value of the function, and those values must actually match the type of the function, which in this case is *bool*. However, what the function does beforehand, and which variables it uses for this, is not specified.

The function itself does the output of the error message in lines 11 to 15 according to the required pattern. In line 16 the user input is queried and in line 18 it is checked whether the letter '*j*' was entered. In this case the function returns the value *true,* in any other case the value *false*.

The *main* function starts at line 27 and is structured exactly like the *error* function. The return type of the function is *int*, the name of the function is *main* and the empty round brackets after the name signal that the function does not take any parameters.

What exactly is to be implemented with this program is not important at this point, so lines 29 and 36 indicate that something is happening there.

In line 30 the variable *goOn* of the type *bool* is created. It is assigned the return value of the *error* function after it has been called with the parameter "*A file could not be ...!*" The function call causes the function to be jumped into first and its code to be executed. Only after the function has been completed does it continue in the main function.

Line 32 checks if the function returned the value *false*, in which case the *main* function would exit and return the value 10 to the caller.

---

[2] A distinction is also made between the *declaration* and *definition* of functions. The *declaration* only creates a function prototype, which is described in Sect. 9.2 on page 100.

Repetition: The caller for *main* is always the operating system. The operating system expects the return value 0 from a program, if the program was terminated properly. Any other value is interpreted as an error message. Of course, the operating system cannot do anything with the returned error, it will just print the value. Therefore, when writing a program, it makes sense to create a documentation of the possible return values in order to be able to interpret the value. In this case, the value 10 corresponds to an abort after a failed file operation. In the end, however, it does not matter which value is used, as long as it is documented what it means.

9.1 Overloading Functions

Occasionally, situations arise where the solution for a task can be solved differently well or efficiently depending on the type of the parameter passed. Sometimes it is also necessary to call a task with different parameter configurations, because the output data can be in different formats, for example as a vector or as a set of coordinates. For these cases C++ has the possibility to overload functions.

This means that several functions may have the same name if only the function parameters within the round brackets differ in number, order or types. The name and the parameter list, where the number of parameters, their order and type are relevant, are called the *signature* of the function. However, two functions do not differ if the parameters are of the same type but have different names, or if two functions differ only in the return type. This information is not part of the *signature*.

An admittedly not very creative but simple example of overloaded functions is given in Listing 9.2. The program generates the following outputs one after the other.

```
This is an int
This is a double
This is a string
```

Listing 9.2 Overloading Functions

```
1    void distinguish(int v)
2    {
3    cout << "This is an int" << endl;
4    }
5
6    void distinguish(double v)
7    {
8        cout << "This is a double" << endl;
9    }
10
11    void distinguish(string v)
```

```
12   {
13        cout << "This is a string" << endl;
14   }
15
16   // The following function cannot be defined ,
17   // since it differs only by the return value of
18   // of the first function.
19   int distinguish(int v)
20   {
21     cout << "This is an int" << endl;
22   }
23
24   int main ()
25   {
26     distinguish (3);
27     distinguish (3.5);
28     distinguish ("t");
29   }
```

9.2   Function Prototypes

Larger programs consist of many different functions, which also frequently call each other. Sooner or later, a problem arises that cannot be solved with previous knowledge. Listing 9.3 shows such a situation in a very simplified way.

Listing 9.3 Reciprocal Function Call

```
1    // Function a needs function b to work.
2    void a()
3    {
4      // ...
5      b();
6      // ...
7    }
8
9    // Function b requires function a to work.
10   void b()
11   {
12     // ...
13     a();
14     // ...
```

```
15   }
16
17   // Main function
18   int main ()
19   {
20      // ...
21      a();
22      // ...
23   }
```

Of course, such a program would cause an error just by having the functions simply call each other in an uncontrolled manner. The program would get stuck and crash almost immediately with an error message. In this example, we will therefore assume that the mutual call is necessary and depends on previously defined conditions.

Nevertheless, a problem arises due to the mutual dependency. A function cannot be used until it has been declared before it is first used. However, since the functions call each other, there is no sequence that would resolve this problem.

What we have learned so far is the so-called *definition* of a function. Here, both the function header and the function body are defined. In addition to the definition, however, it is also possible to *declare* functions. The *declaration* makes it possible to first define the most important information for the function, so that the *compiler* can check that the function has been applied correctly when translating the program. This works even if it has not yet been defined what the function is actually supposed to do.

All important information is summarized in the header of a function. To declare a function, only this information must be specified. Instead of the function body, however, only a semicolon follows in the declaration. The following information must be given in the declaration:

- The return type must be specified so that the translation can check that the context of the function is correct. For example, a function with return type *void* does not produce a value that could be stored in a variable.
- The name of the function is important to recognize the function and to prevent typing errors.
- The parameters must be declared within round brackets. However, it is not necessary to assign names to the parameters for a declaration. The type of the parameters and the correct order are sufficient for a check during translation. However, it is still possible to specify these names already in the declaration.

A function declaration is also called a function prototype. Examples of function prototypes can be found in Listing 9.4.

Listing 9.4 Examples of Function Declarations

```
 1    // A function declaration without specifying the
 2    // Parameter names.
 3    void a(int , double );
 4
 5    // A function declaration with specification of the
 6    // Parameter names. Recommended for maintenance
 7    // and readability reasons.
 8    int b(double x, double y);
 9
10    // Definition of the function a
11    void a(int i, double k)
12    {
13      // ...
14    }
15
16    // Definition of the function b
17    int b(double x, double y)
18    {
19      // ...
20    }
21
22    // Main function
23    int main ()
24    {
25      // ...
26    }
```

After the functions have been declared, they can be used immediately. However, the definition of the function must be done somewhere within the program, otherwise the program cannot be compiled completely. Later programs will consist of several files. We then speak of projects. The function definition can then also take place in other project files.

9.3   References and Arrays as Function Parameters

Functions have the limitation that they have only one fixed return value. The return value behaves rather like the type of a variable. When a function is called, it may calculate a result of a certain type. The return value specifies that type. In the rest of the program, the function call can now be viewed as using a variable of the appropriate type.

Nevertheless, it is often necessary for a function to calculate several values and return them to the calling location. In *C++* this is done via a so-called *call by reference*. This term describes a special notation in the parameter definition of a function within the round brackets. Previously, parameters were always defined by first specifying the variable type and then the variable name. This type of definition is called *Call by Value.*

With *Call by Value,* new variables of the corresponding type are created within the function. If parameters are passed when the function is called, these values are simply copied into the new variables of the function. When the function finishes its task, the return type is copied to the calling location and all function variables are deleted. Since the *call by value* works with local copies of the values, which are deleted at the end of the function call, all changes to the parameter values are lost after the function ends.

Each variable can also be passed to a function with a *call by reference.* To indicate this, an ampersand (&) must be inserted between the type and the parameter name. In this case, no new variable is created. Instead, the function parameter refers to the original variable that was passed to the function. Within the function, the variable can then be addressed with the name of the function parameter, and at the point of the call with the name assigned there. In principle, two names refer to the same variable after a *call by reference.* Of course, it is always possible to use both *Call by Value* and *Call by Reference* when defining the function parameters.

Listing 9.5 shows the definition of function parameters that are passed by *call by reference.* A function prototype was created in the program to illustrate the notation of a *call by reference* in a function declaration.

Listing 9.5 The Exchange of Two Values

```
1    #include <iostream >
2
3    using namespace std;
4
5    // A function declaration for a call by reference.
6    void swap(int&, int &);
7
8    // Main function
9    int main ()
10   {
11     int val1 = 5;
12     int val2 = 10;
13
14     cout << val1 << ", " << val2 << ", ";
15
16     swap(val1 , val2 );
17
18     cout << val1 << ", " << val2;
19
20     // ERROR:
```

```
21      // swap(5, 10);
22      return 0;
23    }
24
25    // Definition of the swap function
26    void swap(int& a, int& b)
27    {
28      int h = a;
29      a = b;
30      b = h;
31    }
```

In line 4 of the program the function prototype for the function *swap* is created. The names of the function parameters do not have to be specified, but it is necessary to write the ampersand after the parameter type to make it clear that this parameter is to use *Call by Reference*. An alternative notation with parameter names is given in line 6.

Inside the *main function* in line 16, the *swap* function is called with the variables *val1* and *val2* previously defined and initialized. After calling the function, the values of the two variables are swapped so that the output of the program is *5, 10, 10, 5*.

An important special case is indicated in line 21. Since the *call by reference* creates a second name for a variable, a variable must exist to which the reference can refer. Consequently, passing constant values is no longer possible for function parameters that use *call by reference*.

In line 26 follows the definition of the function *swap*. It performs a so-called triangular swap. This swap requires an auxiliary variable *h*, in which the value of a variable *a* is temporarily stored. Afterwards, the value in variable *a* is overwritten with the value of variable *b*.Finally, the value cached in *h* can be copied to *b*. After the triangle swap, *a* and *b* have swapped their contents.

The use of *Call by Reference* has another advantage. If only a reference is created instead of a copy, this can of course be done much faster. For this reason, it can make sense to pass a reference even if the value is not to be changed within the function. In this case, the keyword *const* can be used to prevent the parameter from being changed within the function. A function header with a constant reference would be declared by *void f(const int & a);*. However, due to the somewhat confusing definition of the *const* keyword in C++, the function header *void f(int const & a); would* also produce exactly the same result.

9.3.1   Arrays as Parameters

Arrays are also allowed as function parameters. Unlike other variables, however, arrays are always passed by reference. It is therefore not possible to change the values of the array within the function without this change also occurring at the calling point.

Listing 9.6 shows declarations and definitions of functions that take one-dimensional arrays as parameters. In addition to the normal option of omitting the names of the parameters in the declaration, there is also the option of specifying the size of the array (as in the case of the function *f* in Listing 9.5), or omitting it (as in the case of the function *g*). However, there is no difference in the result between these two variants. Since *C++* does not check the size of an array and it is even possible to pass smaller or larger arrays to the function, this specification can also lead to confusion.

Listing 9.6 Functions with Arrays

```
1    // Function declarations with arrays.
2    void f(int [5]); void g(int []);
3
4    // Main function.
5    int main ().
6    {
7      int values [5];
8
9      f(values);
10     g(values);
11   }.
12
13   // Definition of the function f.
14   void f(int w[5]).
15   {.
16     //...
17   }.
18   // Definition of the function g.
19   void g(int w[]).
20   {.
21     //...
22   }.
```

If it is to be ensured that the values of an array cannot be changed, the keyword *const* can also be used here. The function declaration *void f(const int w[]);* or *void f(int const w[]);* would prevent values of the array from being changed within the function.

For multi-dimensional arrays, the size of the array can only be omitted for the first dimension. *C++ is* otherwise unable to resolve the coordinates of the array correctly. A declaration of a function that takes a multidimensional array of size 5 · 5 as a parameter could consequently be *void f(int w[][5]);* or *void f(int w[5][5]);*.

9.4   Advanced: Pre-assigned Function Parameters

Functions can significantly simplify programs and by the possibility to pass parameters, they can additionally adapt their behavior individually to the needs of the respective situation. However, it can happen that the various functions should actually behave almost always the same and should only do something different in exceptional cases.

An example could be a function that is to write a series of hyphens as a separator line in the console. The separator line should always consist of twenty hyphens, but for a single output the length of the separator line must be only ten hyphens. For this case *C++* offers the possibility to assign default values to function parameters.

Listing 9.7 defines a *separator* function that implements the previously described example. In the function prototype, the parameter *b* is preset to the value 20. This default value must not be repeated in the function definition. Only if no function prototype exists, the preassignment must be made in the function definition.

Listing 9.7 Predefined Parameters for Functions

```
1    #include <iostream >
2
3    using namespace std;
4
5    // Function declarations with preassignment.
6    void separator(int b = 20);
7
8    // Main function
9    int main ()
10   {
11     separator ();
12     separator (10);
13   }
14
15   // Definition of the separator function
16   void separator(int b)
17   {
18     for (int i=0;i<b;i++)
19     {
20       cout << "-";
21     }
22     cout << endl;
23   }
```

When the *separator* function is called, the parameter no longer needs to be specified. In this case the value 20 is set automatically. If a value other than 20 is to be used, the function can be called as usual by specifying the parameter value.

A special situation arises when several parameters are to be preassigned. Here it must be ensured that preassigned parameters must always be at the end of the parameter list. Since preassigned parameters can be omitted when calling, it would otherwise not always be possible to assign the subsequent parameters correctly. There is also a special feature in the function call. If the value of the last parameter is to be changed, values must still be passed for all other parameters in this case, even if they are already preassigned. Listing 9.8 shows some examples.

Listing 9.8 Examples of Preassigned Function Parameters

```
1    // Function declarations
2
3    // OK
4    void test1(int a = 0, int b = 1);
5
6    // Not OK
7    void test2(int x = 0, int y = 0, double r);
8
9    // OK
10   void test3(int x, int y, double r = 1.0);
11
12   // OK
13   void test4(int x, int y = 0, double r = 1.0);
14
15   // OK
16   void test5(int x = 0, int y = 0, double r = 1.0);
17
18   // Function calls
19   test1 (5);        // OK a = 5, b = 1
20   test5 (3);        // OK , if x should be 3
21                     // Not OK , if r = 3.0 is required
22   test5(0, 0, 3.0) // OK x = 0, y = 0, r = 3.0
```

9.5   Advanced: Variadic Functions

In some rare cases, it may be necessary to write a function whose number of parameters cannot be specified in advance. Such a function is called a variadic function. An example of this is the *printf* function, which produces output to the console. The first parameter of the *printf* function is a *char* field that specifies a text into which values are to be inserted at certain positions. The places where values are to be inserted are marked by predefined special characters in the text. The values to be inserted then follow as additional parameters when the function is called. The function call.

```
printf("%i values were created", 5);
```

would, for example, produce the output *5 values were created* on the console. In this case, the predefined special character *%i* means that a parameter is to be inserted into the text as an integer.

Of course, with such a function it is not clear how many parameters are to be inserted into the text. Specifying the number or types of parameters would result in an unrealizable number of combinations, all of which would have to be covered by individual functions. To prevent this, *C++* offers the possibility of defining functions that have a variable number of parameters.

Although this is a very helpful offering of *C++*, it is still not easy to use and error prone because of the many unknowns.

Listing 9.9 implements a variadic function that is to calculate the sum of *n* numbers. In order to be able to develop a function with a variable number of parameters, the library *stdarg.h* must be included, since the functions required for this are otherwise not available. Even with these functions, *C++* offers no way of finding out how many parameters have been passed to the function. For this reason, it makes sense to query this information with the first parameter, which is defined as an *int*. This is followed by three items in the parameter list to make it clear that any number of additional parameters can follow here. The function prototype and the function definition header hardly differ in these functions.

Listing 9.9 An Example of a Variadic Function

```
1    #include <iostream >
2    #include <stdarg.h>
3
4    using namespace std;
5
6    // Function prototype
7    double sum(int n, ...);
8
9    // Definition of a variadic function
10   double sum(int n, ...)
11   {
12     double result = 0;
13
14     va_list parameterList;
15     va_start(parameterList , n);
16
```

```
17      for (int i = 0; i<n; i++)
18      {
19        double summand = va_arg(parameterList , double);
20        result += summand;
21      }
22      va_end(parameterList);
23
24      return result;
25    }
26
27    // Main function
28    int main ()
29    {
30      cout << sum(5, 1.0, 2.0, 3.0, 4.0, 5.0);
31    }
```

To be able to store a parameter list, a variable of the type *va_list* is required. In the current example, the variable is called *parameterList*. This variable must be initialized using the *va_start* function. The function requires two parameters, firstly the parameter list to be initialized and secondly the variable *n*. However, the variable *n* does not specify how many parameters are to be taken over, but is the name of the parameter after which the variable parameter list is to begin. It is important to de-initialize the parameter list using the *va_end* function before exiting the function.

In order to calculate the sum of the passed parameters in the example program, each individual parameter must be evaluated. The function *va_arg* always returns the next parameter from the parameter list that is passed to the function. In addition, the function must be told how the parameter is to be interpreted. In our example the value should be interpreted as *double*. This is another source of error: It is not possible to determine the actual variable type. The example program only works if the passed values can actually be interpreted as *double*. Additionally, not all variable types can be used, *char*, *int* and *double* work, but *float* does not.

To get the *n* parameters that were passed, the function is called within a *for loop*. Again, C++ does not check whether the number of parameters passed actually corresponds to the value *n*.

Within the main function the function is called with some test values. The output of the program is 15 as expected.

9.6   Advanced: Recursive Calls

The basic principle of recursion is very simple to understand at first glance. An operation is recursive if the same rules are applied over and over again to a data set. In C++, recursive functions are, in the simplest case, those that call themselves over and over again. In more complicated cases, multiple functions can call each other to solve a problem recursively. Listing 9.3 was already such a recursive example. However, to understand in detail what recursion means and to solve your own problems recursively, a deeper understanding and some experience is needed.

First of all, it is important to understand that every function that is called stores a set of information in memory, called the *stack*, or stack memory. The name for this memory was cleverly chosen because new information is always placed "on top" of the stack and only this information can be easily retrieved. There is also talk of "LIFO" in this context. This means *Last In First Out,* the last information that was put down is the first to be removed. Just like a stack of plates.

A function's information includes the return value, the function parameters, the function's local variables, and the return address, which indicates where the program must jump when the function has finished. This is done anew for each function call, even if a single function calls itself over and over again. So even though a function is called the same thing, new variables are created on the stack for each new call, and they can have individual values.

Another insight is that this memory is limited. Thus, there cannot be an infinite number of function calls in succession. On the contrary, a function that calls itself without limit will cause a crash practically immediately.

So why should a problem be solved recursively if it is difficult to understand and there is also the risk of a memory-induced crash? In fact, simple problems such as factorial computation or Fibonacci numbers, which at first sight invite to be solved recursively, are often very inefficient if a recursive approach is chosen. These problems can be solved much more efficiently using a loop. This approach is called iterative.

However, for more complex problems, such as sorting numbers, very efficient recursive solutions can be found, such as the *Quicksort* algorithm. There are also recursive data structures that are commonly used and can be searched very efficiently using recursive approaches.

To get started, however, it is useful to begin with a simple example, even if it is not efficient: The factorial of a number n *is* calculated by multiplying all numbers from 1 to n together. From this, a recursive rule can be derived:

$$n! = n \cdot (n-1)!, \quad with \quad 1! = 1 \tag{9.1}$$

The factorial of a number n can be calculated by multiplying n by the factorial of the number $n - 1$, where the factorial of the number 1 is defined as 1. This rule translates directly into the recursive Listing 9.10.

Listing 9.10 Recursive Solution of the Faculty Calculation

```
1    #include <iostream >
2
3    using namespace std;
4
5    // recursive faculty calculation
6    unsigned int faculty(unsigned int n)
7    {
8      if (n == 1) return 1;
9
10     return n * faculty(n - 1);
11   }
12
13   // Main function
14   int main ()
15   {
16     cout << faculty (5) << endl;
17   }
```

The function *faculty* recursively calculates the factorial of the number n. It is important to note that there is a non-recursive path in addition to the recursive path that calls the faculty function again. This non-recursive path is chosen if the number n passed has a value of 1, since that value is already fixed. It ensures that the function does not make any further recursive calls after a finite number of calls.

If the value of the variable n does not equal 1, the function calculates the result of the product of n and the recursive call to the function *faculty* with the parameter $n - 1$. This function calls itself with the parameter $n - 2$ and so on until the parameter eventually equals 1. After that, each function in turn returns the result of its calculations, until at the end the factorial is completely calculated.

An iterative solution to the faculty problem is presented in Listing 9.11. By using a loop, the problem can be solved much more efficiently because it requires much less memory. The iterative solution does require three *int* variables to compute the result. However, the recursive solution requires one *int* variable for each function call.

Listing 9.11 Iterative Solution of the Faculty Calculation

```
1    #include <iostream >
2
3    using namespace std;
4
```

```
5    // iterative faculty calculation
6    unsigned int faculty(unsigned int n)
7    {
8      unsigned int result = n;
9
10     for (int i = n-1;i>1;i--)
11     {
12       result *= i;
13     }
14
15     return result;
16   }
17
18   // Main function
19   int main ()
20   {
21     cout << faculty (5) << endl;
22   }
```

9.7   Advanced: *static*

Within functions, the keyword *static* has a different meaning than with global variables. If a variable is created and initialized as a static variable within a function, this variable is not created anew with each function call and deleted after the function is completed. Instead, the variable is created and initialized only once, when the function is called for the first time.[3] From that point on, the variable exists and is not deleted until the program terminates.

Static variables can be used to store state within functions that is passed from function call to function call. For example, a static variable can be used to count how many times a particular function has been called. Listing 9.12 implements the *callCount* function.

Listing 9.12 Using Static Variables to Count Function Calls

```
1    #include <iostream >
2
3    using namespace std;
4
5    void callCount ()
```

---

[3] In fact, the initialization of static variables is not limited to function bodies. The rules explained here apply to all static variables that are created inside blocks, i.e., inside curly braces.

```
6    {
7        static int count = 0;
8        count ++;
9
10       cout << "Function call: " << count << endl;
11   }
12
13   // Main function
14   int main ()
15   {
16       for (int i = 0; i < 10; i++)
17       {
18           callCount ();
19       }
20   }
```

Within the function, the static variable *count is* created first and initialized with 0. This initialization is located within the function, but is only called once during the first function call and is ignored thereafter.

With each function call, the value of the variable is incremented by 1 and printed to the console.

Within the main function, the *callCount* function is called 10 times in a row within a loop. The output of the program is:

```
Function call: 1
Function call: 2
Function call: 3
Function call: 4
Function call: 5
Function call: 6
Function call: 7
Function call: 8
Function call: 9
Function call: 10
```

So the static variable actually retains its value between function calls, counting the number of times the function is called.

Exercises

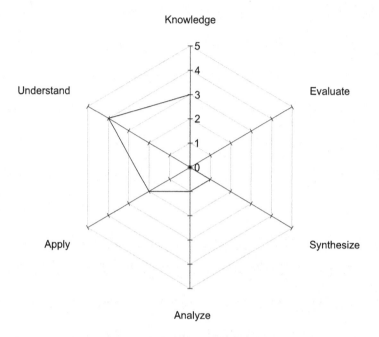

Network diagram for the self-assessment of this chapter

9.1   Function Prototype

List the information needed to create a functional prototype.

9.2   Return Value

Write down the function header of a function f that has no return type. Three variables are to be passed as parameters, which have the types *int, double,* and *char.* You are to indicate the function body by curly brackets.

9.3   *Call by Reference*

What is meant by the term *Call by Reference*?

9.4   **Variadic Functions**

Explain in your own words what properties a variadic function has!

9.5   **Recursion**

What is the difference between a recursive and an iterative solution?

9.6   *static*

Explain the meaning of the term *static* in the context of functions!

9.7   **Function Overloads**

What is meant by "overloaded functions"?

9.8   **Input and Output Functions**

Write a program in which two functions are used to perform the inputs and outputs respectively.

To do this, first implement an *input* function that is to read in text from the console and return it to the caller as a return value. Use the *getline(cin, text);* statement to read in the text, since the operator ⟩⟩ would interrupt the input at the spaces.

The second function, *output*, is to receive the text and output it to the screen. There should be a row of hyphens above and below the text, which is exactly as long as the output text.

Within the main function, the *input function* is to be called first and the result is to be stored in a variable of the type *string*. Before the output function is called, all spaces ('') in the input text are to be replaced by asterisks ('*').

9.9   Recursion

Design a program that calls the *recursion* function in the *main* function and passes it the parameter 0.

(a) Now write the function *recursion,* which takes a parameter c of type *int.* It is to call itself with the parameter $c + 1$ if the value of c is less than 100. After the call has been made, the value of c is to be printed to the screen.
(b) Modify the program so that the output of the variable c occurs before the recursive call. Explain the modified output!

9.10   Program Analysis

Analyze the following program. To do this, try to find out what the individual program lines do in terms of content and deduce the task of the program.

Commands have been used that you don't know yet. Try to research them!

Do not type the program, but try to understand what is happening without assistance!

```
1    #include <iostream >
2
3    using namespace std;
4
5    int func(int val[], int s, int e)
6    {
7      if ((e - s) == 0) return val[s];
8
9      int h = (e + s) / 2;
10     int e1 = func(val , s, h);
11     int e2 = func(val, h + 1, e);
12
```

```
13      return e1 + e2;
14   }
15
16   int main ()
17   {
18     const int N = 100;
19     int values[N];
20
21     for (int i = 0; i < N; i++)
22     {
23       values[i] = i + 1;
24     }
25
26     cout << "Result: " << func(values , 0, N - 1)
27          << endl;
28
29     return 0;
30   }
```

9.11   Output of Parameters of a Variadic Function

In this task, you are to develop the variadic function *myPrint,* which takes one parameter of type *string* and then allows the user to pass any number of additional parameters.

Within the function, the *string* is to be run through and printed to the console character by character. Whenever there is an asterisk ('*') in the function, however, not the asterisk is to be printed, but one of the additional parameters, which is always to be interpreted as *int.*

The main function should call *myPrint* with the parameters *("– + –*– + –*– + –*",* *1, 2, 3).*

Classes and Structures

10

Short and Sweet

- A class describes a blueprint for objects.
- It creates a new self-defined data type (the *string* data type is a good example).
- An object, or instance, is a concrete manifestation of a class (e.g., a *string* variable that stores a specific text).
- In addition to data, a class can have functions.
- There are a number of specialized functions in a class:
  - The constructor initializes an object of a class
  - The destructor de-initializes an object of a class
  - The operators allow, among other things, the definition of mathematical connections, comparisons or relations ($+, -, *, /, =, = =, [], \ldots$).

So far in this book, programs have been presented that consist of individual instructions that are grouped into functions. This type of programming is called procedural programming, since functions are often also called procedures in computer science. At the beginning of the nineties of the last century, however, the term object-oriented programming appeared.

The idea of object-oriented programming is to combine data and functions that belong together in terms of content in a common structure, the class. The class serves as a blueprint for objects that can actually store and manipulate data. This type of programming allows a new approach to the structure of a program and a different idea of data structures.

An illustrative example that can be used for a class is a vector in a two-dimensional coordinate system. First, it must be determined what data make up such a vector. For the vector, this is obviously the two coordinates x and y.

B. Tolg, *Computer science to the Point*, https://doi.org/10.1007/978-3-658-38443-2_10

Every vector in a two-dimensional coordinate system has these two properties. In addition, further information can be derived from these two properties, such as the length of the vector, or its angle to the x-axis. Furthermore, there are a number of mathematical functions defined for vectors, such as addition, subtraction or scalar product, as well as relations, such as the equality of two vectors.

In procedural programming, the data and the functions that actually belong to it would have no relation to each other. Different coordinate pairs can be stored in a multidimensional array, while functions on this array would take over the arithmetic operations.

With the help of a class, however, a new data type *Vector2D* can be defined, which combines both the data and the associated functions in a single structure. When programming with classes, a distinction is made between functions and variables "inside" and "outside" of classes. Functions and variables that are declared as part of the class when the class is declared are called member functions, or *methods,* and member variables, or *attributes*. Especially with member variables, it is good practice to indicate by the name of the variable that it is a variable within the class. For this reason, all member variables in this book are preceded by "m_".

There are different ways where classes can be placed within a program. In principle, it is possible to write all classes and the main function in a single file. However, this quickly becomes confusing even for smaller projects. It is also possible to write almost all functions of the class within the class declaration. However, the *C++* language offers the possibility to split classes into two independent files, which can then be included in other files. This notation is used consistently in this book, so that example programs now always consist of at least three parts. Each class has a *header* and a *cpp* file. In addition, there is the main program, which is located in its own *cpp* file.

Most development environments allow you to automatically generate the files needed for a class. Mostly, the declaration and the implementation of the class is also prepared automatically.

The class declaration is located in the file *Vector2D.h,* a so-called *header* file. In this book, *header* files have been used several times, always in connection with the *#include* statement. The *header* files contain the class declaration, or more graphically, a sort of table of contents for the class. In the class declaration, all variables and functions that the class should have are declared. The example Listing 10.1 shows the *header* file for the class *Vector2D*.

Listing 10.1 The Vector2D Class (Vector2D.h)

```
1    // Include -Guard
2    #ifndef _VECTOR2D_H
3    #define _VECTOR2D_H
4
5    // Class declaration
6    class Vector2D
```

```
7    {
8    public:
9       // Variable declaration
10      double m_x;
11      double m_y;
12   };
13
14   // End of the Include -Guard
15   #endif // _VECTOR2D_H
```

The *header* file starts with some preprocessor commands called *include-guards*, which perform an important task. They prevent a class from being declared more than once. This mechanism is necessary because every file in which the class will be used later must have an *#include* statement that includes the *Vector2D.h header* file. When the preprocessor encounters an *#include* statement, it copies the contents of the specified file to the location of the *#include* statement. Consequently, if there were no *include guard,* every *#include* statement would attempt to declare the class again. This would already fail on the second attempt.

So how exactly does the *include guard* work? The first statement *#ifndef* stands for *if not defined.* This statement checks if the following text _VECTOR2D_H has not been defined yet. When the preprocessor encounters this statement for the first time, the text is of course not defined yet.[1] In this case, the statement is correct and the following text is processed unchanged. The second statement that follows is then the *#define* statement, which ensures that the corresponding text is defined immediately. So for all following *#include* statements the *#ifndef statement* will fail and ignore the complete text up to the *#endif statement.*

Some preprocessors support the non-standard *#pragma once* command, which also has the properties of an *include guard.* In this case, the *header* file must only start with this command to prevent multiple declarations.

The actual class declaration begins with the *class* statement followed by the name of the class. Within curly brackets, the functions and variables can then be declared. The class declaration is concluded with a semicolon.

The first statement within the class is a visibility level keyword, a very important concept in object-oriented programming: data encapsulation. This concept puts the responsibility for the data completely in the hands of the class. In C++, this means that when you program a class, you can specify which data can and cannot be read and modified from outside the class. There are three so-called *visibility levels:*

---

[1] Since, among other things, terms can be defined by included libraries, it is important that unambiguous expressions are chosen.

- *public:* The keyword *public* specifies that these attributes and methods should be publicly accessible. They can be used by functions outside the class as well as by member functions within the class.
- *protected:* Variables and functions declared after the keyword *protected* are protected from external access. They may only be used by member functions declared within the class.
- *private:* Variables and functions declared after the keyword *private* initially behave as if they had been declared after the keyword *protected*. External access is prevented; only member functions of the class may access these elements. In addition, elements declared after the *private* keyword are not inherited. Inheritance of class properties is another important concept in object-oriented programming, and is explained in more detail in Sect. 10.4.

The different visibility levels can be used as often as desired during a class declaration. A visibility level is valid until it is replaced by another one. If no visibility level is specified for a class, the level is *private*. This means that each element of a class has a visibility level. To avoid errors, however, you should always specify the visibility level explicitly.

The concept of data encapsulation allows you to declare helper functions and member variables that can only be used within the class. As long as a single person is working on a program, this may not seem like a good idea, since that person can access everything at any time anyway. And also knows the program perfectly. But if more than one person is working on a project, the classes of the others are comparable to a black box, whose content is not necessarily known in detail. If then the call of a helper function has to be prepared, or the call of the function alone would not produce a meaningful result, it is important to protect this helper function. Instead, a public member function should be provided that does all the preparation, or performs the calls in the correct order.

In the case of member variables, a decision must be made as to whether the value range of the variable is restricted or whether other variables may be affected by the change of a variable value. If either of these applies to a member variable, access from outside the class should be prevented. Instead, public auxiliary functions should be written to allow access.

The current example deals with a vector in a two-dimensional coordinate system. The class has only two member variables m_x and m_y, whose value range is not restricted and whose values are independent of each other. Therefore, there is no reason to protect these member variables from access. The declaration of the member variables is done as already known, by first specifying the data type followed by the name of the variable. The statement is terminated with a semicolon.

The second file shown in Listing 10.2 is the so-called *cpp* file, in which the member functions of the class are implemented. Since there are currently no member functions, this file is almost empty for now. The *#include* statement makes sure that the class declaration is known in the *cpp* file. Until now, *#include* statements always used angle brackets to specify the filename. Whether brackets or quotes are used depends on where to look for the files. Angle brackets are used to search paths defined by the development environment or the *compiler*. They are always used when standard libraries are included. With quotes the search radius is increased, the directory where the file with the *#include* statement is located and sometimes others are searched as well. The quotation marks are always used when *header* files of the own project are to be included.

Listing 10.2 The Vector2D class (Vector2D.cpp)

```
1 #include "Vector2D.h"
```

To obtain an executable program, there must still be a main function, which in this case should be in the file *Project.cpp* and is shown in Listing 10.3.

Listing 10.3 The Main Program (Project.cpp)

```
1     #include <iostream >
2     #include "Vector2D.h"
3
4     using namespace std;
5
6     // Main function
7     int main ()
8     {
9        Vector2D v1;
10
11       // ...
12
13       // Read in values
14       cin >> v1.m_x;
15       cin >> v1.m_y;
16    }
```

To use the new class, the header file of the class *Vector2D* must be included using the *#include* statement. Within the main function an object of the class *Vector2D* is now created by a variable definition and named *v1*. Since the member variables m_x and m_y of the class *Vector2D* have been declared as *public*, access is also possible from within the main function. To access the member variables of an object, the name of the object must first be written, in this example *v1*, followed by a dot and the name of the member variable to be accessed. The dot serves as a "door opener" into the object.

In principle, m_x and m_y within the main program now behave like normal variables with a slightly longer name. However, these variables are assigned to object *v1*. It would now be easy to create a second object *v2*, which would in turn have two member variables. Since both objects belong to the class *Vector2D*, it is easier to think of the objects as vectors and work with them. With two arrays of type *double* it would be more difficult to see the connection or to interpret four values as two vectors.

10.1   Constructors and Destructors

In Chap. 5 of this book it is already recommended to always initialize variables. This advice also makes sense for classes. Since only the class has access to its variables in any case, there must therefore be a way for a class to perform variable initialization. But it is not only variables that need to be initialized. More complex classes may require more elaborate configurations to be made when a new object is created. It may also be important to de-initialize elements when an object of a class is deleted.

To achieve this, the *C++* language has two types of functions, the constructors and the destructor. It is certain that the first function called for a new object is a constructor. Similarly, it is certain that the last function called before an object is deleted is the destructor. Both of these functions are called only once in the life cycle of an object. There can be multiple constructors that allow an object to be initialized in different ways. However, there is always only one destructor.

Listing 10.4 adds some constructors and a destructor to the previous example. Both the constructors and the destructor have some special features regarding the return type and the name. Both the constructors and the destructor have no return type, not even *void*. The name of a constructor always corresponds exactly to the class name. If there are multiple constructors, the normal function overloading rules introduced in Sect. 9.1 apply. All constructors must have different parameter configurations. The name of a destructor is also the same as the class name, but the name is always preceded by a tilde (~).

Listing 10.4 Constructors and Destructor of the Vector2D Class (Vector2D.h)

```
1    // Include -Guard
2    #ifndef _VECTOR2D_H
3    #define _VECTOR2D_H
4
5    // Class declaration
6    class Vector2D
7    {
8    public:
9       // Standard constructor
10      Vector2D ();
11      // various constructors
12      Vector2D(double x, double y);
13      // Conversion constructor
14      explicit Vector2D(double l);
15      // Copy constructor
16      Vector2D(const Vector2D &v);
17
18      // Destructor
19      ~Vector2D ();
```

```
20
21     double m_x;
22     double m_y;
23   };
24
25   // End of the Include -Guard
26   #endif // _VECTOR2D_H
```

In this example it makes sense to offer several constructors. It may be that a new vector is to be created without further configuration. In this case, the constructor that does not take any parameters is used. This constructor is called the default constructor and exists even if it is not defined. In fact, it was already used in Listing 10.3 when object *v1* was created. However, this automatically generated variant of the default constructor does not perform variable initialization. Therefore, it still makes sense to implement the default constructor yourself, since this is the only way to initialize your own variables in a meaningful way. The example in Listing 10.3 also makes it clear that constructors are called differently than other functions. Although only the variable type and the variable name are specified, this already corresponds to a constructor call without the round brackets indicating a function call.

In some cases it is easier if an object can be created right away with the correct configuration. In the case of a vector, the position could already be available in Cartesian coordinates. For this reason the second constructor was created.

Conversion constructors are always needed when it should be possible to convert a value or an object of a certain type into another. They always have exactly one parameter of the type that is to be converted into an object of the class. In this case, a single *double* value is to be interpreted as a vector that is parallel to the x-axis and has length *l*. The C++ language will now use this constructor to automatically convert values of type *double* to a *Vector2D* in a function call, for example, should that be necessary. Such calls and conversions are called *implicit* calls. It may be that such conversion constructors are needed, but *implicit* calls are not desired by the C++ language or even cause errors. For this case the keyword *explicit* exists. It ensures that C++ cannot call the constructor automatically, but that a conversion must be written explicitly in the program.

The explicit conversion is done by a type conversion, or a *typecast.*

The fourth constructor is the so-called *copy constructor.* The *copy constructor* is also generated automatically, even if it was not explicitly defined. Its task is very simple, it creates an exact copy of an object. Normally, this task is very easy to solve by simply copying the complete memory area containing an object to the position of the new object. The size of an object is known from the class declaration and the internal structure of the objects is also always identical. However, there can be major problems if custom memory is reserved, as then the automatic copy will no longer work. This is explained in more detail in Sect. 11.8. However, the basic structure of a *copy constructor* should be explained here.

The *copy constructor* has a very important task in the *C++* language. Whenever a parameter is passed or a return value is returned in a function call, a copy of the corresponding variable must be created. If the variable is an object of a class, the *copy constructor* is always called without this being explicitly specified in the program.

A *copy constructor* always has a single parameter. This parameter always has the type of the class itself and must be passed by reference. If the parameter were not passed by reference, a copy of the object would have to be created when the *copy constructor* function is called. However, this would require the *copy constructor*, which is just being defined.

With newer *compilers*, it must also be ensured that the value passed cannot be changed by marking the reference as constant with the keyword *const*.

The destructor does not perform any task in the current example and was only declared for the sake of completeness.

A definition must now be implemented for each of the constructors and the destructor. This is done in the *cpp* file shown in Listing 10.5.

Listing 10.5 Constructors and Destructor of the Vector2D Class (Vector2D.cpp)

```
1    #include "Vector2D.h"
2
3    // Standard constructor
4    Vector2D :: Vector2D ()
5    // Initialization
6    : m_x (0.0)
7    , m_y (0.0)
8    {
9      // Value assignment
10     m_x = 0.0;
11     m_y = 0.0;
12   }
13
14   // various constructors
15   Vector2D :: Vector2D(double x, double y)
16   : m_x(x)
17   , m_y(y)
18   {
19   }
20
21   // Conversion constructor
22   /* explicit */ Vector2D :: Vector2D(double l)
23   : m_x(1.0)
24   , m_y (0.0)
25   {
26   }
```

```
27
28    // Copy constructor
29    Vector2D :: Vector2D(const Vector2D &v)
30    : m_x(v.m_x)
31    , m_y(v.m_y)
32    {
33    }
34
35    // Destructor
36    Vector2D ::~ Vector2D ()
37    {
38    }
```

The first special feature is found directly in the function name of the constructors. To make clear that a function is defined here, which was declared within the class *Vector2D*, the name of the class must be prefixed with two colons.

Since each object is created by exactly one of the constructors, a variable initialization must take place in each of the constructors, which uses the passed function parameters. There are two ways to assign values to the variables. First, it is possible to assign a value within the function body of a constructor. In the standard constructor of program 10.5, such a value assignment was made for all parameters as an example. At first, this seems to be the intuitive way, but it has the disadvantage that constant values of the class cannot be assigned values this way. This only works for the initialization, which is regulated for constructors by a special notation.

To initialize variables in a constructor, a colon can be written after the parameter list of the function to start the initialization. This is followed by the names of the variables to be initialized, separated by commas, and the initial values in round brackets. This notation is called *initialization list*.

Since no parameters are passed with the standard constructor, the variables can be initialized with freely selectable values. Without information, it is best to create a null vector.

In the next constructor, two parameters for the x and y coordinates are passed. Here we can see why it makes sense, among other things, to provide the names of the member variables with an $m\_$. It is now clear that the local function parameters are called x and y and the member variables of the class are called m_x and m_y. Within the constructor, the member variables must now only be initialized with the values of the matching function parameters.

The conversion constructor should transform a *double* value l into a vector parallel to the x-axis with length l. This is done very easily by writing the value l directly into the member variable m_x, while m_y is given the value 0. It is important to note at this point that the keyword *explicit* must not be repeated again in the function definition. Nevertheless, in order to point out that this constructor may only be called explicitly, the keyword was inserted as a comment in the header.

The *copy constructor* has only one parameter *v*, which is of the type of the self-written class *Vector2D* and is passed by reference. As already shown in the example Listing 10.3, the member variables of a class object can be accessed using a point. Since both the passed *Vector2D* and the new object have the same member variables, all member variables of *v* can thus be used to initialize their corresponding member variables in the new object. There is only one peculiarity here. Since the *copy constructor* is part of the class, it is always considered a member function of the class. So even if two objects of the class are being processed here, the *copy constructor* always has access to all member variables. Even if they have been declared with the visibility level *protected* or *private*.

The destructor is defined for the sake of completeness. In this example, however, it will not fulfill any function and therefore remains empty.

In the main function of the program 10.6 the different constructors can now be tested.

The first four objects created in the program use the constructors in the order they were created in the class. First the default constructor, then the constructor that takes Cartesian coordinates, followed by the conversion constructor and the copy constructor. These calls are relatively obvious, except for the fact that no parentheses need to be appended when the default constructor is called.

It is less obvious that a copy of the vector *v2* is created when the function *output* is called. Here C++ automatically calls the *copy constructor*.

The value assignment *v3* = 5 actually does not work because the conversion constructor has been marked *explicit*. If the *explicit* keyword in Listing 10.4 were omitted, the line would work because the conversion constructor could then be called implicitly.

The subsequent value assignment *v3* = *Vector2D*(5), on the other hand, always works because the conversion constructor is called explicitly (Listing 10.6).

Listing 10.6 The Main Program (Project.cpp)

```
1    #include <iostream >
2    #include "Vector2D.h"
3
4    using namespace std;
5
6    void output(Vector2D v);
7
8    // Main function
9    int main ()
10   {
11     Vector2D v1; // Standard constructor
12     Vector2D v2(2.1, 3.2); // Constructor
13     Vector2D v3 (2.3); // Conversion constructor
14     Vector2D v4(v2); // Copy constructor
15
16     output(v2);
```

```
17
18     v3 = 5; // Works only , without explicit
19     v3 = Vector2D (5);
20     // ...
21   }
22
23   void output(Vector2D v)
24   {
25     //...
26   }
```

10.2   Member Functions

Further values are to be derived from the member variables of the vector. For example, the length or angle of the vector can be calculated in polar coordinates. In addition, the possible value range of the vectors is restricted to the interval from −10 to 10 on both coordinate axes. Thus it is now necessary to protect the member variables of the class from external access.

In general, it is good practice to prefix the names of functions that return values from the class with the prefix *get*. Similarly, function names of functions that set values within the class should start with the prefix *set*. This is not mandatory by any means, but it makes it easier to keep track.

Listing 10.7 shows the modified class declaration. In addition to the constructors and destructor, a number of function prototypes have been added. The function prototypes behave exactly as described in Sect. 9.2. Each function that returns a value has been given a name beginning with the prefix *get*. The function that changes the values of the vector starts with the prefix *set*. The notation in which the first word of a name is written in lower case and then each subsequent word is started with an uppercase letter is called *camel case* notation. It is a naming convention commonly used in computer science.

Listing 10.7 Member Functions of the Vector2D Class (Vector2D.h)

```
1    // Include -Guard
2    #ifndef _VECTOR2D_H
3    #define _VECTOR2D_H
4
5    // Class declaration
6    class Vector2D
7    {
8    public:
9      // Standard constructor
10     Vector2D ();
```

```
11      // various constructors
12      Vector2D(double x, double y);
13      // Conversion constructor
14      explicit Vector2D(double l);
15      // Copy constructor
16      Vector2D(Vector2D &v);
17
18      // Destructor
19      ~Vector2D ();
20
21      // Member functions
22      double getAngle ();
23      double getLength ();
24      double getX ();
25      double getY ();
26
27      void setCartesian(double x, double y);
28
29   protected:
30      double m_x;
31      double m_y;
32   };
33
34   // End of the Include -Guard
35   #endif // _VECTOR2D_H
```

The member variables *m_x* and *m_y* have now been declared as *protected* to prevent direct access to the variables from outside the class. The access is now indirect. The *getX* or *setCartesian* function can be accessed from outside the class because they have been declared as *public*. These member functions of the class in turn have access to the protected member variables.[2]

This seems to be cumbersome at first sight, but this approach has a big advantage. Within the functions, it can be checked whether the values that are to be set are within the permitted limits. This allows the class to ensure that the values stored in an object of the class always comply with the rules.

Of course, it is possible to extend the class with many useful member functions. However, the functions shown are sufficient for this example.

---

[2] Some of my students have problems at the beginning to understand why these functions have no parameters. The background is that these functions are always applied to an object of the class. The information they are supposed to reflect or change is already present in the object. Nevertheless, there may of course be situations in which additional information needs to be transported via the parameters, even in the case of a class function. This works in exactly the same way as for all functions.

The implementation of the member functions is done in the *cpp* file of the class and shown in programs 10.8 and 10.9. Since some mathematical functions are used in the new functions, a new library must be included with *#include*, the *cmath* library. It allows the use of functions like *sin* or *cos* for sine and cosine, *sqrt*, which stands for *square root,* and many more.[3]

The default constructor remains unchanged in this example because it always sets the member variables to 0, which is within the allowed interval. However, the constructor that takes two coordinates *x* and *y* must now check whether the values are within the interval. Since the *setCartesian(…)* function must also check the bounds, it makes sense to call the function directly in the constructor. This way, the checking of the interval bounds only needs to be done in the *setCartesian(…)* function. On the one hand this saves a duplicate implementation, but on the other hand it also avoids errors. In larger programs, duplicate implementations can easily result in errors being fixed in only one place, while others are forgotten. Awareness of this should be developed so that such situations are avoided (Listing 10.8).

Listing 10.8 Modifications to the Constructors of the Vector2D Class (Vector2D.cpp)

```
1    #include "Vector2D.h"
2    #include <cmath >
3
4    // Definition of a constant for Pi
5    const double PI = 3.141592653589793238462643383279 5;
6
7    // Standard constructor
8    // ...
9
10   // different constructors
11   Vector2D :: Vector2D(double x, double y)
12   {
13     setCartesian(x, y);
14   }
15
16   // Conversion constructor
17   /* explicit */ Vector2D :: Vector2D(double l)
18   {
19     setCartesian(l, 0);
20   }
21
```

---

[3] In some implementations, mathematical constants are also defined, but not in all. Sometimes these constants cannot be accessed directly, a specific expression must be defined before the *cmath* library can be included. Often this expression is *#define _USE_MATH_DEFINES*. However, since this is not standardized, this book refrained from using one of the variants to give a general introduction to C++. Instead, a constant *PI* was defined and used.

```
22    // Copy constructor
23    // ...
24
25    // Destructor
26    // ...
27
28    // ...
```

The conversion constructor must now also observe the specified interval limits, so the same solution was chosen here. Listing 10.9 now shows the new member functions of the class.

Listing 10.9 Member Functions of the Vector2D Class (Vector2D.cpp)

```
1     // ...
2
3     double Vector2D :: getAngle ()
4     {
5       return atan2(m_y , m_x) * 180 / PI;
6     }
7
8     double Vector2D :: getLength ()
9     {
10      return sqrt(m_x * m_x + m_y * m_y);
11    }
12
13    double Vector2D ::getX ()
14    {
15      return m_x;
16    }
17
18    double Vector2D ::getY ()
19    {
20      return m_x;
21    }
22
23    void Vector2D :: setCartesian(double x, double y)
24    {
25      if (x > 10) x = 10;
26      if (x < -10) x = -10;
27      if (y > 10) y = 10;
28      if (y < -10) y = -10;
29
30      m_x = x;
31      m_y = y;
```

32   }

The function *getAngle(…)* calculates the angle of the vector to the positive x-axis using a function called *atan2(…)*. This function, just like the *atan(…)* function, calculates the arc tangent, but differs in the function parameters. While the function *atan(…)* takes only one parameter, which is calculated from *m _ y/m _ x*, the function *atan2(…)* takes the values *m_y* and *m_x* in this order in two separate parameters. The background is that the *atan(…)* function can only distinguish between two quadrants of the coordinate system because of the two signs. For the *atan2(…)* function, four different sign combinations can arise, which allow to distinguish between all four quadrants of the coordinate system. A mathematical explanation for all formulas used here can be found in Papula (2014).

With the help of the function *getLength(…)* the length of the vector can be calculated by the Pythagorean theorem. The length thus also corresponds to the radius required for the polar coordinates.

To query the current values of the member variables, the functions *getX()* and *getY()* can now be used. Since this is the only task of these functions, the implementation is very short. Since values of the function must be stored in memory for each function call, this means that a small amount of additional time is required to organize the function call. For very small functions, of course, the organizational overhead is proportionally much greater than for functions in which a lot happens. The *C++* language therefore offers the possibility of using the clarity of a function without suffering a loss of speed. For very small functions, the keyword *inline* can be prepended to the class declaration in the *header* file. This keyword, like *explicit*, is also prefixed only in the declaration, not in the definition. In this case, the *compiler* can decide whether to make a function call or to copy the function's code directly to the call location. In the *header* file, the program line would then be *inline double getX();*, or *inline double getY();*.

The function *setCartesian(…)* now ensures with some *if-statements* that the values with which the member variables are initialized are always in the interval from −10 to 10.

In the main function in Listing 10.10, the member functions can be tested. Since the member variables are now *protected*, direct access is no longer possible. Two auxiliary variables *x* and *y* are needed to read in the values.[4]

Listing 10.10 The Main Program (Project.cpp)

```
1     #include <iostream >
2     #include "Vector2D.h"
3
4     using namespace std;
5
```

---

[4]This is a common procedure for more complex programs that have a graphical user interface, since the input is often not read in the format that will be processed later.

```
6    // Main function
7    int main ()
8    {
9      Vector2D v1; // Standard constructor
10
11     // Auxiliary variables
12     double x = 0.0;
13     double y = 0.0;
14
15     // Output
16     // Value input
17     cin >> x;
18     cin >> y;
19
20     v1.setCartesian(x, y);
21
22     cout << "Vector length: " << v1.getLength () << endl;
23   }
```

With the help of the function *setCartesian(…)* the read values can then be passed to the vector. The resulting vector length can be calculated using the *getLength( )* function.

10.3   Operators

For classes that represent constructs from mathematics, it is important that the mathematical operations defined for these constructs can also be realized in the classes. Of course, it is always possible to write functions that perform the operations, but it would be much more intuitive if the normal mathematical operations could also be performed using the usual notation. To realize this, there are special functions for classes in the *C++* language that differ from other functions both in terms of programming and usage. They make it possible to implement mathematical operations in the usual notation. This special type of function is called an operator.

In addition to the mathematical operations, there are other possible applications for operators that make it possible to reinterpret language elements of *C++* for your own classes. When operators are implemented for their own classes, this is called operator overloading. A non-exhaustive list of operators that can be overloaded in *C++ is* shown in Table 10.1.

To illustrate how operators work, the *Vector2D* class will now be extended. In Listing 10.11, the declarations for the operators have been added to the *header* file. Parts of the program have been replaced by comments this time to highlight the new parts of the program. These parts of the program are unchanged from Listing 10.7.

Table 10.1 Overloadable operators of the language C++

Type	Operators
Arithmetic operators	+, −, *, /, %, ++, −−, Sign: +, −
Assignment operators	=, +=, −=, *=, /=, %=, &=, l=, ^=, <<=, >>=
Comparison operators	==, !=, >, <, >=, <=
Logical operators	!, &&, ll
Bit operators	~, &, l, ^, <<, >>
Field operators	[]

Listing 10.11 Operators of the Vector2D Class (Vector2D.h)

```
1    // Include -Guard
2    #include <iostream >
3
4    using namespace std;
5
6    // Class declaration
7    class Vector2D
8    {
9    public:
10     // ...
11
12     // Array operator
13     double operator [] ( int n);
14     // Scalar multiplication
15     Vector2D operator *( double right );
16     // Scalar product
17     double operator *( Vector2D right );
18
19     // external operators
20     friend ostream &operator <<( ostream &out , Vector2D );
21     friend Vector2D operator *( double left , Vector2D right );
22
23    protected:
24     // ...
25    };
26
27    // Output operator
28    ostream &operator <<( ostream &out , Vector2D right );
29
30    // Scalar multiplication
31    Vector2D operator *( double left , Vector2D right );
32
33    // End of the Include -Guard
```

What makes operators confusing, especially for beginners, is the fact that operators can be declared in two different ways. In addition, for some of the operators only one of the two declaration options is selectable. Why this is so will be explained a little later. First of all, it must be understood how an operator basically works.

The names and calls of operators work differently from other functions to mimic the typical notation of mathematical functions. To make this clear, it is easier to start with the operators that are declared outside of classes. As an example, let's define a multiplication of two vectors whose result should be the scalar product. The exact mathematical explanation of what a scalar product is can be found in Papula (2014). For this example, only the formula 10.1 is important for the time being.

$$\vec{a} \cdot \vec{b} = \begin{pmatrix} a_x \\ a_y \end{pmatrix} \cdot \begin{pmatrix} b_x \\ b_y \end{pmatrix} = a_x \cdot b_x + a_y \cdot b_y = c \tag{10.1}$$

Now there are three questions two be answered:

- Which variable type can be used to describe the result of the multiplication?
- What type of variable is on the left side of the multiplication sign?
- What type of variable is on the right side of the multiplication sign?

For this example, these questions can be easily answered by looking at the formula. To the left and right of the multiplication sign is a vector and the result of the multiplication is a real number. The vectors are best represented by the class *Vector2D*, which is being developed here, and for the real number a variable of type *double* is a good choice. These answers can be translated directly into a declaration for an operator. The basic structure of a global operator, i.e. an operator declared outside a class, is:

```
Return type operator* (typeL nameL, typeR nameR);
```

In this specific example, it can be translated into:

```
double operator* (Vector2D left, Vector2D right);
```

Calling such an operator in another function is now very untypical to other functions:

Listing 10.12 Calling an Operator

```
1    Vector2D left (1 ,0);
2    Vector2D right (2 ,2);
3
4    double returnvalue = left * right;
```

So when an operator is called, the return value is used normally. However, the name of the operator is not *operator** when it is called, but just *. The parameters are not written in parentheses after the function call, but without parentheses to the left and right of the operator. The first parameter of the declaration always corresponds to the value to the left of the operator and the second parameter always corresponds to the value to the right of the operator.

If the same operator is now declared within a class, the first parameter is always automatically of the type of the class itself. The declaration within the class would be

```
double operator* (Vector2D right);
```

The implementation of the two operators is also slightly different, as Listing 10.12 shows using the example of a scalar multiplication. However, the function call of the operator is still the same as in Listing 10.11, except that the left parameter must now be of type *Vector2D*. In fact, the *compiler* even recognizes from the left parameter in which class to search for a declaration of the corresponding multiplication.

Now, in mathematics, there are operations that always have the same result, regardless of the order of the parameters. These operations are called commutative. However, this is not true for all operations. Therefore it is important that C++ offers the possibility to distinguish between the different parameter orders. Since when an operator is declared inside a class, the left parameter, which must also be left when called, is always of the type of the class, some operators must be declared outside the class. There is, in fact, another restriction to consider. An operator that is declared inside a class only makes sense if one of the two parameters is of the type of the class. However, this cannot be guaranteed for operators where both parameters can be freely chosen. Therefore, it is consistent to declare them outside of classes.

But now another problem arises. If an operator is declared outside a class, then it has no access to the protected elements of the class. However, this would be desirable if, for example, mathematical operations are involved where only a different sequence should be realized. Therefore C++ offers the possibility to declare functions and classes within a class, which should be allowed access to protected elements. The keyword for this exception is *friend*. Within the program 10.11 there is an additional line within the class for all operators declared outside the class. This line consists of the keyword *friend* and a function prototype. Since these prototypes are unique, this allows you to specify the functions that are to be "friends" of the class.

For the *friend* declaration, the visibility level is irrelevant. The result is identical whether the declaration is made in a *public, protected,* or *private* realm. The "friendship" is also not transitive. This means that the "friends" of a friendly class do not automatically become "friends" of their own class. Additionally, "friendship" cannot be inherited. If a class inherits from a friendly class, a "friendship" does not automatically exist with its own class. What inheritance means exactly is explained in Sect. 10.4 and is not important here.

Listing 10.11 shows an example of five different operators, three of which are declared within the class. The first operator is an array operator and makes it possible to write square brackets after an object of the class, as if it were array with two elements.

The second operator is to enable a scalar multiplication, i.e. a product of a vector with a scalar, as in formula 10.2. Since the operator was declared within the class, the left parameter must therefore always be of type *Vector2D*. Thus, this operator could be used to enable the arithmetic operation vector times scalar, but the reverse notation scalar times vector could not. Listing 10.13 shows both cases in the application.

Listing 10.13 Calling Scalar Multiplication

```
1   Vector2D left (1 ,0);
2
3   // OK
4   Vector2D returnvalue = left * 5;
5
6   // Not OK
7   Vector2D returnvalue2 = 5 * left;
```

$$\vec{a} \cdot \lambda = \begin{pmatrix} a_x \\ a_y \end{pmatrix} \cdot \lambda = \begin{pmatrix} a_x \cdot \lambda \\ a_y \cdot \lambda \end{pmatrix} = \vec{b} \tag{10.2}$$

To enable the second case as well, an external operator *double operator* (double left, Vector2D right);* was defined, which has a scalar as first parameter and a vector as second parameter.

For the third operator, which is to realize the scalar product shown in formula 10.1, there is only one declaration, because both parameters are vectors and therefore the order does not matter.

A special operator is the output operator, which must also be declared outside of a class. It is based on the class *ostream* whose object *cout* has already been used several times. With the help of the output operator, it should be possible to print an object of the class *Vector2D* directly through *cout*. This is not possible so far, because of course nobody defined how to deal with a class *Vector2D* when implementing *cout*. With the help of this operator, however, this can be done.

In Listing 10.14 now follows the implementation of the operators. Here, too, the already known functions were replaced by three dots, since no changes were made to the previous programs.

Listing 10.14 Implementing the Operators of the Vector2D Class (Vector2D.cpp)

```
1   #include "Vector2D.h"
2   #include <cmath >
3   // ...
4
5   // Array operator
6   double Vector2D :: operator [](int n)
7   {
8     if (n == 0) return m_x;
9     if (n == 1) return m_y;
```

```
10
11     return nan ("");
12   }
13
14   // Scalar multiplication
15   Vector2D Vector2D :: operator *( double right)
16   {
17     Vector2D result;
18
19     result.m_x = m_x * right;
20     result.m_y = m_y * right;
21
22     return result;
23   }
24
25   // Scalar product
26   double Vector2D :: operator *( Vector2D right)
27   {
28     return m_x * right.m_x + m_y * right.m_y;
29   }
30
31   // Output operator
32   ostream &operator <<(ostream &out , Vector2D right)
33   {
34   out << "Vector2D (" << right.m_x
35       << ", " << right.m_y << ")";
36
37     return out;
38   }
39
40   // Scalar multiplication
41   Vector2D operator *( double left , Vector2D right)
42   {
43     return right * left;
44   }
```

It is noticeable that only the operators, whose declaration takes place within the class, were supplemented by the prefix *Vector2D::*. The operators declared outside the class are global functions, which must not get this addition, because they are not part of the class.

The field operator is invoked by adding square brackets with an index after the name of an object. So an object *v1* of class *Vector2D* could be used like an array by the array operator. The following lines:

```
Vector2D v1(3.2, 1.5);

cout << v1[0] << " : " << v1[1] << " : " << v1[2] << endl;
```

shall produce the output:

```
3.2 : 1.5 : nan
```

Here *nan* stands for *not a number* and is intended to make clear that there is no element of the vector for this index. The operator itself only checks the value of the parameter *n*. For *n* = 0 the value of *m_x is returned* and for *n* = 1 the value of *m_y*. In any other case, the value *nan* is returned using the *nan("")* function provided by *C++*.

The operator for scalar multiplication corresponds to the typical structure of mathematical operators. First, a variable of the return value type is created to store the result. In the second step, the values for the result are calculated. In this case, by multiplying the member variables of the class *m_x* and *m_y* by the parameter *r*. Since the multiplication was declared within the class and therefore has only one function parameter, the left parameter of the operator is always the object of the class itself. Finally, the calculated result value only has to be returned as a result.

The scalar product is calculated in a very similar way. However, since the calculation of the result value is so short, no additional variable is created, but the result is calculated directly after the *return* statement. Again, the left parameter of the operation is the object of the class, so the member variables *m_x* and *m_y* are used directly. The right parameter *r* was passed as a function parameter. Since the operator is part of the class *Vector2D*, the protected member variables can be accessed directly, so that the calculation can be performed according to formula 10.1.

The output operator << has several meanings in the *C++* language. For numbers, such as *int,* which can be interpreted as binary numbers, it is used to shift all bits of the number a certain number of places to the left. However, for classes, the operator is also used for printing with the help of the *ostream* class. If the operator is used for the latter, the operator must follow the rules of the *ostream* class. A stream first collects information and usually releases it in blocks. If an output operator is to be written, an object of the class *ostream* must be passed as a reference as the first parameter, so that when it is written *cout< <v1;*, the first parameter, in this case *cout,* is changed by the call. In addition, the passed object must also be returned, so the return type must also be *ostream&*, so that a concatenated notation, as *cout< <v1 < <endl*; works.

Within the operator, the *ostream* reference, called out in this example, can be used in the same way as *cout*. Output can be produced using data types known to *ostream,* such as *string, int,* or *double*. Finally, after the output has been made as desired, the *ostream* object must be returned with the *return* statement.

The last operator for scalar multiplication was written to be able to write the scalar to the first position in a multiplication as well. However, the result of both notations is identical, because the multiplication of a scalar with a vector is commutative. Again, the arithmetic operation should not be implemented twice, because this could lead to an error later. So the operator internally only reverses the order of the parameters and returns the result of the already implemented operator as return value.

In the main function of Listing 10.15, the new operators can now be tested.

Listing 10.15 The Main Program (Project.cpp)

```
1    #include <iostream >
2    #include "Vector2D.h"
3
4    using namespace std;
5
6    // Main function
7    int main ()
8    {
9       // Variable definition and initialization
10      Vector2D v1(1, 0);
11      Vector2D v2(2, 2);
12
13      // Output operator
14      cout << "v1: " << v1 << endl
15            << "v2: " << v2 << endl;
16      // Scalar product
17      cout << "Scalar product: " << v1 * v2 << endl;
18      // Scalar multiplication
19      cout << "5 * v1: " << 5 * v1 << endl;
20      cout << "v1 * 5: " << v1 * 5 << endl;
21      // Array operator
22      cout << "v1.m_x: " << v1[0] << endl;
23      cout << "v1.m_y: " << v1[1] << endl;
24      cout << "Error: " << v1[2] << endl;
25   }
```

The main function begins by first creating two vectors. The first vector runs parallel to the x-axis and has the length 1, while the second vector is at a $45°$ angle to the first vector with the length $\sqrt{8} \approx 2.828$. The first thing to test is the output operator. Without the operator, it would not be permissible to write an object of class *Vector2D* directly into a *cout* statement. However, with the help of the operator, the output can be implemented this way. The complete output of the program is:

```
v1: Vector2D(1, 0)
v2: Vector2D(2, 2)
Scalar product: 2
5 * v1: Vector2D(5, 0)
v1 * 5: Vector2D(5, 0)
v1.m_x: 1
v1.m_y: 0
Error: nan
```

where the first two lines are generated by the output operator.

The scalar product calculates a result of 2. This value can be checked by calculating the arc cosine of the quotient of the scalar product and the product of the two vector lengths. The result is the angle between the two vectors and its magnitude is as expected $acos\left(2/\left(1 \cdot \sqrt{8}\right)\right) = 45°$.

To test scalar multiplication, the vector $v1$ is multiplied once by the number 5 from each side, increasing its length by a factor of 5 in both cases.

Finally, the individual elements of the vector $v1$ can now be obtained using square brackets, as if $v1$ were an array. If an index is specified that is neither 0 nor 1, *nan is* returned as the error value.

To be able to implement further operators, it is absolutely necessary to have a basic understanding of pointers. For this reason, the implementation of further operators is continued in Sect. 11.8.2.

10.4   Inheritance and Polymorphism

An important concept of object-oriented programming is the ability of classes to inherit their properties to other classes. These "inheritors" then have the ability to add new properties and functions to the class without changing the original class, called the base class. But that's not all. It is even possible to replace functions that exist in the base class with new functions with the same name and different functionality. This concept is called polymorphism.

If two classes have the same member functions, this alone does not indicate that the two classes should inherit from each other. As an example, we will mention two classes that can draw a triangle or a circle in the console. Both classes have identical functions, they can calculate the area and the perimeter and both classes have a function *paint,* with which either a triangle or a circle is drawn. Inheritance makes no sense here, because both classes have little relation to each other and store completely different data.[5]

---

[5] In this case, however, it may be useful to define a class as an interface from which both classes can inherit.

What does inheritance mean in object-oriented programming and especially in *C++*? When a class inherits from another class, it automatically receives all member variables and member functions that were declared and defined in the other class within the visibility levels *public* or *protected*. However, all attributes and methods from the visibility level *private* are not accessible to the inheriting class. Therefore, if a class *A* inherits from a class *B* that only uses the visibility levels *public* or *protected*, *A* can immediately do everything that *B* can do. This concept is particularly useful if class *A* is an extension of class *B* in terms of content, and many of class *B*'s properties can actually be adopted in a meaningful way.

The great strength of this system is that *C++* remembers that class *A* possesses certain properties of *B*. If a reference of type *B* is now expected in a function call, an object of class *A* can also be passed because of the inheritance. Within the function, any property that *B* offers can then be used. However, the behavior depends on the class of the object that was passed.

This description is very theoretical and abstract, and will be made concrete in an example in a moment. However, the basic idea that the type of a passed variable using inheritance does not always correspond to the type expected by the function should be kept in mind.

Particularly good examples of inheritance are systems that need to manage various display windows, buttons, text fields, and more, since these very classes reuse many properties of the other windows. Each window has a certain width and height, reacts somehow to the attempt to move it and much more. Unfortunately, this example in particular is very complex and not suitable for beginners.

A less complex example results from a classification system that is to record the students and staff of a university. Here, it makes sense to collect the basic data that must be stored for both students and staff in a class *person*, from which the other two classes *student* and *staff* can inherit.

In Listing 10.16, the class *person* is declared. The *include-guards* have been shortened by comments in the following, but are to be implemented analogously to the previous examples. Of course, much more information and functionality could be implemented in the class, but since the topic here is inheritance, the example will be kept small. First of all, with this class, it makes no sense to provide a default constructor, since an entry for a person of which nothing is known is unnecessary. So in this example, you can only create people whose data is completely known.

Listing 10.16 Declaration of the Class *person* (person.h)

```
1    // Include -Guard
2    #include <string >
3
4    using namespace std;
5
```

```
 6   // Class declaration
 7   class person
 8   {
 9   public:
10      // Constructor
11      person(string firstName , string lastName );
12      // Destructor
13      virtual ~person ();
14
15      virtual void print ();
16
17      string getFirstName ();
18      string getLastName ();
19   protected:
20      // Name
21      string m_firstName;
22      string m_lastName;
23   };
24   // End of the include guard
```

When declaring the destructor, a new keyword is used: *virtual*. This keyword identifies functions that may be redefined by the inheriting class. For classes in which at least one function is declared as virtual, a table is created, the so-called *virtual function table* or *vtable*. In this table *C++* remembers which functions were implemented by the current class and which were not.

It was mentioned earlier that the type of the passed variable does not always have to match the expected type when inheritance is involved. With the help of the table, *C++* can remember which function to use even if the type of the variable does not match expectations. If the *virtual* keyword is missing, functions in inheriting classes can still be reimplemented. However, in this case the entry in the *vtable* is missing. If an object from an inheriting class is now passed to a function that expects a variable of the type of the base class, the function of the base class is called.

The behavior is similar for destructors. If the destructor is virtual, an object can be deleted even if the type of the object is not known. In this case, the *vtable* can be used to check which destructor is the correct one.

The *print* function serves as an example of polymorphism and is to be overridden in the inheriting classes, therefore it was also declared as *virtual*. The two *get* functions are intended to allow the first and last name of the person to be queried individually. These functions are to be inherited unchanged, without being able to be changed.

The implementation of the class *person* is shown in Listing 10.17. As with other keywords, *virtual* is specified only in the declaration and is not repeated again in the definition. In the example program, the keyword was nevertheless written as a comment before the function. This is a good practice, which simplifies the work with complex programs, since it serves the clarity. The constructor initializes the member variables using the passed parameters and the destructor has no task in this example either.

Listing 10.17 Implementation of the *person* Class (person.cpp)

```
 1   #include "person.h"
 2   #include <iostream >
 3
 4   using namespace std;
 5
 6   person :: person(string firstName , string lastName)
 7   : m_firstName(firstName)
 8   , m_lastName(lastName)
 9   {
10   }
11
12   /* virtual */ person ::~ person ()
13   {
14   }
15
16   /* virtual */ void person :: print ()
17   {
18     cout << "person (" << m_firstName
19            << " << m_lastName << ")" << endl;
20   }
21
22   string person :: getFirstName ()
23   {
24     return m_firstName;
25   }
26
27   string person :: getLastName ()
28   {
29     return m_lastName
30   }
```

The *print* function prints the name of the class in the console, as well as the first and last name within round brackets. The functions *getFirstName* and *getLastName* return only the content of the corresponding member variable.

Listing 10.18 now shows the class *student,* which is to inherit and add all properties of the class *person.* First of all, the *header* file *of* the class *person* must be included using an *#include* statement, otherwise the class *person* would not be known. By this inclusion also the datatype *string* is automatically known, because it was also included in the *header* file of *person.*

Listing 10.18 Declaration of the Class *student* (student.h)

```
1    // Include -Guard
2    #include "person.h"
3
4    using namespace std;
5
6    // Class declaration
7    class student : public person
8    {
9    public:
10     // Constructor
11     student(string firstName , string lastName
12             , int matriculationNumber );
13     // Destructor
14     virtual ~student ();
15
16     virtual void print ();
17
18     int getMatriculationNumber ();
19   protected:
20     // Matriculation number
21     int m_matriculationNumber;
22   };
23   // End of the Include -Guard
```

When declaring the class *student*, it must now be made clear that the class is to inherit from *person*. This is done by adding a colon after the keyword *class* and the name of the class, followed by a visibility level and the name of the base class to inherit from. This is followed by the normal class declaration.

The visibility levels for inheritance are also *public, protected*, and *private,* although *public* is almost always used as the visibility level for inheritance. With the visibility level *protected,* all elements of the visibility levels *public* and *protected* of the base class are inherited, but are all assigned the visibility level *protected* in the inheriting class. This means that none of the elements can be accessed from outside. The same happens with the visibility level *private,* except that all elements receive the visibility level *private* and would therefore no longer be inherited. Additionally, classes that inherit *private* from a base class can no longer be used as objects of the base class, as with the other visibility levels. This is a severe restriction of the possible functionality, which is only needed in the rarest of cases.

All member variables of the class *person* are now also available in the class *student*, so first and last names can already be stored. However, students have an additional matriculation number, which a person does not have. For this reason, an additional member variable *m_matriculationNumber of* type *int* is declared.

The constructor of the class *student* should again only be callable if all information about the students is available, so that it also contains the matriculation number as a function parameter in addition to the first and last name.

Since the member functions of the class *person* were also inherited, the class *student* already has the functions *getFirstname* and *getLastname,* so that these do not have to be declared again. In addition, the class *student* should have the function *getMatriculationNumber*, with which the matriculation number can be queried.

The implementation of the *student* class is shown in Listing 10.19. And the first peculiarity occurs right away in the constructor. The variables *m_firstName* and *m_lastName* were declared only in the base class *person,* but not in the inheriting class *student.* For this reason, the two variables must not be initialized during initialization, that is, after the colon and before the body of the constructor. In general, it can happen that inheritance is from an unknown class, so that it is not clear what actually happens in the constructor of the base class. It is therefore advisable to rely on the base class constructor when initializing the base class variables. The constructor of the base class may be called during initialization. In this case, the values of the *firstName* and *lastName* parameters are simply passed through. The constructor does the initialization. Within the class *student* only the initialization of the member variable *m_matriculationNumber* has to be added.

Listing 10.19 Implementation of the *student* Class (student.cpp)

```
 1    #include <iostream >
 2    #include "student.h"
 3
 4    using namespace std;
 5
 6    student :: student(string firstName , string lastName
 7               , int matriculationNumber)
 8    : person(firstName , lastName)
 9    , m_matriculationNumber(matriculationNumber)
10    {
11    }
12
13
14    /* virtual */ student ::~ student ()
15    {
16    }
17
18    /* virtual */ void student :: print ()
19    {
20      cout << "Student (" << m_firstName << " "
21           << m_lastName
22           << ", Matriculation number: "
```

```
23                 << m_matriculationNumber << ")" << endl;
24     }
25
26    int student :: getMatriculationNumber ()
27    {
28      return m_matriculationNumber;
29    }
```

Generally, in an inheritance hierarchy, if a class has inherited from other classes, the constructor of the base class is always called first. After that, each constructor follows the path of the hierarchy, up to the constructor of the class whose object is currently being created. For the destructors, the path is the other way around. Here, the destructor of the current class is called first, and the destructor of the base class is called last.

The function *print* has been newly implemented for the students. It now prints the name and the matriculation number. Occasionally it can be useful if the functions of the base class can be used without having to rewrite everything in the inheriting class. The C++ language therefore allows access to the *members* of the base class in an inheriting class, even if they are overwritten by your own functions. To allow access, the name of the base class must first be written, followed by two colons[6] and the name of the element from the base class. The function *print* could also have been implemented as follows:

```
/*virtual*/ void student::print()
{
  person::print();
  cout << ", matriculation number: "
<< m_matriculationNumber << endl;
  }
```

In this implementation, the *person::print();* statement would first produce the output through the *print* function of the *person* class. After the closing parenthesis of the output, a new line would then begin and the output of the matriculation number would follow there.

The *getMatriculationNumber* function returns the value of the member variable, analogous to the other *get* functions.

To test the classes and their functions, a function *testInherit* was written, which is called by the main function in Listing 10.20. First, the classes *person* and *student* are made known by including their *header* files so that they can be used.

Listing 10.20 The Main Program (Project.cpp)

```
1    #include "person.h"
2    #include "student.h"
```

---

[6]This is also used in the *cpp* files to make it clear that a function belongs to a particular class.

```
3
4    void testInherit(person &p)
5    {
6      p.print ();
7    }
8
9    // Main function
10   int main ()
11   {
12     person p(" Maxima", "Sampleman ");
13     student s("Max", "Samplewife", 1234567);
14
15     testInherit(p);
16     testInherit(s);
17   }
```

The function *testInherit* takes a reference of type *person* as parameter. Within the function nothing more is done than calling the *print* function of the passed object. If the parameter in line 4 is passed as a *call by reference*, objects of class *student* can also be passed. Since these are still the original objects, the modified functions of the class *student* are also called in this case. If the object were passed by *Call by Value*, a function call with the class *student* would still work, but the object would be copied to an object of the class *person*. The result would always be a call to the function *print* of the class *person*.[7]

In the main function, a variable *p* of the type *person* and a variable *s* of the type *student* are created with sample values. Then the function *testInherit* is called with both variables. The output of the program is:

```
Person(Maxima Sampleman)
Student(Max Samplewife, matriculation number: 1234567)
```

So the concept of polymorphism has been successfully applied here. Although the function expects an object of the class *person* in both cases, the combination of inheritance and passing by reference allows an output to take place here that is dependent on the class passed.

If the alternative *print* function of the class *student* was used, which shares the implementation of the base class, the output would change.

```
Person(Maxima Sampleman)
Person(Max Samplewife)
, matriculation number: 1234567
```

---

[7] With other classes, however, various problems can occur with this procedure. It would therefore be advisable to define the type conversion exactly in the class.

In this case, the *student* class would first produce the output defined in the *person* class, and then the supplementary output from the *student* class.

10.5   Advanced: Abstract Classes

If inheritance hierarchies are implemented, in which many classes inherit from each other, it is often useful to define interfaces that specify exactly which functions must be implemented and which must not. Often, however, these interface classes have no task of their own, so that on the one hand there is no meaningful implementation for the functions and, in addition, it should be prevented that objects of this class can be created at all. In *C++*, such classes are called *abstract*.

An example of such a situation are the already mentioned graphic objects, such as line, polygon or circle. All these objects should have a common interface that allows to calculate properties, such as the area or the perimeter, and to draw the objects. In *C++* it is possible to declare a class that defines this interface. If all graphic objects then inherit from this interface class, it is ensured that all graphic objects adhere to the same rules.

The declaration of interfaces has another advantage. When a class is implemented that will later work with the graphic objects, this class only needs to know the interface and not all possible classes that inherit from the interface at some point. Since all inheriting classes must always adhere to the interface, it is ensured that the processing works in any case.

Listing 10.21 shows the declaration of an interface class for the graphics objects. Again, only a few functions have been implemented to make the concept clear.

Listing 10.21 Declaration of an Abstract Class (IgraphicObject.h)

```
1    // Include -Guard
2    #include <string >
3
4    using namespace std;
5
6    // Class declaration
7    class IgraphicObject
8    {
9    public:
10      // Constructor
11    IgraphicObject ();
12    // Destructor
13    virtual ~IgraphicObject ();
14
15    virtual void print () = 0;
16    virtual double getArea () = 0;
17    virtual double getPerimeter () = 0;
18
```

```
19      string getClassName ();
20   protected:
21      // Class designation
22      int m_className;
23   };
24   // End of the include guard
```

It has proven to be good practice to identify interface classes by their names. One way to do this is to prefix the name of the class with a capital *I* for *interface*. Of course, this is not mandatory and it has no effect on the class. However, it makes it easier to keep track of things in large projects.

The constructor and destructor are implemented normally for abstract classes, even if the tasks are usually limited to initializing and deinitializing some variables. Of course, it depends on the concrete situation, but since an abstract class is created as the base class of an inheritance, it makes sense to think about declaring the destructor as virtual and do this in case of doubt.

To make the class abstract, functions of the class must be declared as *virtual* and then set to 0. Such functions are called purely virtual. The =0; at the end of a function declaration means that no implementation exists for this function within the class. In fact, an implementation may still exist, but only the class inheriting from the base class may refer to it.

The very existence of a single purely virtual function ensures that no objects of the class may be created and that the class is therefore abstract. In this example, the function *getClassName* was added to illustrate this. There can still be any number of functions that are fully implemented, but the class is still abstract.

Listing 10.22 implements the constructor, destructor, and *getClassName* function of the abstract class.

Listing 10.22 Implementing an Abstract Class (IgraphicObject.cpp)

```
1    #include "IgraphicObject.h"
2
3    IgraphicObject :: IgraphicObject ()
4    : m_className (" IgraphicObject ")
5    {
6    }
7
8    /* virtual */ IgraphicObject ::~ IgraphicObject ()
9    {
10   }
11
12   string getClassName ()
13   {
14      return m_className;
15   }
```

The constructor simply initializes the member variable *m_className* with the name of the class, which can be queried by the function *getClassName*. Such a construction often proves to be useful when output of the program is to be created in log files during debugging. This way it can always be traced which class is responsible for a process. Of course, inheriting classes must adapt the content of the member variable *m_className* accordingly for the system to work.

Definitions for purely virtual functions could still be added to the implementation if there are implementations that can be reused in many inheriting classes. However, this is not necessary and only makes sense in a few cases.

10.6   Advanced: Structures

The *C++* language is an extension of the *C* language and has adopted its language elements. In *C*, it was already possible to create more complex data types by combining several variables using the *struct* keyword. Listing 10.23 shows an example of such a structure.

Listing 10.23 An Example of a *struct* Data Structure

```
1    struct Container {
2        int id;
3        double value;
4    }
```

Confusion often arises about when to use a *struct* and when to use a *class* in *C++*, and what the difference actually is between these two constructs. In the *C++* language, the difference is actually minimal. With a *class*, if nothing else is specified, the visibility level is *private*. Even with inheritance, a class would be inherited *private* if nothing else is specified. With a *struct*, the visibility level is public in both cases without further specification. Otherwise, the two constructs behave identically. Everything else discussed in the previous chapters can be applied to classes as well as to structures.

So what's the point of the two constructs, isn't it enough to use classes? In principle, yes.

However, the two constructs can be used to indicate different types of classes. If a construct is to be developed that has constructors, operators and functions, a class should be used. It represents the idea of object-oriented programming with all the associated possibilities.

Occasionally, however, it may be necessary to design a data container that is intended to bundle some data for a specific application purpose, but does not require any functions or operators. Access to all elements of this container should be possible directly, at best a constructor and destructor are helpful. In these cases, many programmers use a *struct*.

10.7   Advanced: *const* and *static*

The two keywords *const* and *static* have already been introduced elsewhere in this book. In the context of classes, however, some special features are added, which will be introduced in this chapter.

10.7.1   *const*

If constant objects of a class are created, C++ must be able to recognize which functions change the state of the object and which do not. A *get* function that only returns the content of a variable could be called without problems, but a *set* function could not. However, since *get* and *set* have no meaning for C++, this is not a working criterion.

Functions that are also to be usable with constant objects must be explicitly marked by the keyword *const*. This procedure is called *const correctness*. In Listing 10.24, the *Vector2D* class is modified so that functions that do not change the state of the class can also be called on constant objects.

Listing 10.24 Declaring the Vector2D Class Using *const correctness* (Vector2D.h)

```
1    // Include -Guard
2    // Class declaration
3    class Vector2D
4    {
5    public:
6       // Constructors and Destructors
7
8       // Member functions
9       double getAngle () const;
10      double getLength () const;
11      double getX () const;
12      double getY () const;
13
14      void setCartesian(double x, double y);
15
16   protected:
17      // Variable declaration
18   };
19   // End of the Include -Guard
```

The constructors and destructor of the class are obviously functions that change the state of the object, as variables are initialized or deinitialized. But all *get* functions can be complemented by the keyword *const,* since they do not change the respective object.

However, the *set(...)* function allows the state of the object to be changed, so the *const* keyword must not be added to this function.

Unlike other keywords such as *inline* or *virtual, const* must also be specified when implementing the respective function. This is shown as an example in Listing 10.25.

Listing 10.25 Member Functions of the Vector2D Class Considering *const correctness* (Vector2D.cpp)

```
1    // ...
2    double Vector2D :: getAngle () const
3    {
4      return atan2(m_y , m_x) * 180 / PI;
5    }
6    // ...
7    void Vector2D :: setCartesian(double x, double y)
8    {
9      //...
10     m_x = x;
11     m_y = y;
12   }
```

Classes can additionally contain member variables that have been marked as constant. In contrast to variables, these constants can only be assigned a value by the initialization list in the constructor. A value assignment is neither allowed in the constructor, nor in the rest of the class.

10.7.2 *static*

Until now, all member variables and member functions described in this book have been bound to an object. This means that although the class declares functions and variables, an object must be created in order to actually use the variables and functions.

However, there can also be data and functions that can be assigned to a specific class but are independent of individual objects. An example of this is a count variable that is to count all objects of a certain class. In principle, this is quite simple: Each time an object of a class is created or destroyed, a variable must be incremented or decremented by one. But this only works if this variable is the same variable for all objects of the class. And that's exactly what the keyword *static* makes possible.

It is similar for functions declared as *static*. The functions can be called without an object of the class existing. Since static functions belong to the class and not to a specific object, only static variables of the class can be used within a static function. However, you can pass function parameters as normal.

Listing 10.26 shows the declaration of a class whose only job is to count how many objects of the class exist.

Listing 10.26 Declaration of the Class *counter* (counter.h)

```
1    // Include -Guard
2    // Class declaration
3    class counter
4    {
5    public:
6       // Constructor
7       counter ();
8       // Destructor
9       ~counter ();
10
11      static int getCount ();
12
13   protected:
14      static int m_count;
15   };
16   // End of the Include -Guard
```

The constructor and destructor have been declared normally. The class should have a static variable *m_count,* which should count the number of objects of the class. In addition, there should be a function *getCount*, which should also be static, and with whose help the value of the variable *m_count* can be queried.

Static variables cannot be initialized in the constructor of a class like normal variables. The *compiler* prevents this, but even if not, it would not make sense. If static variables were initialized in the constructor, this would happen with every new object and that would go against the idea of a static variable. Instead, static variables are initialized like global variables. To make it clear that they are variables of a class, the name of the variable must be preceded by the name of the class followed by two colons. In Listing 10.27, this is implemented immediately after the *#include* statement.

Listing 10.27 Implementation of the *counter* Class (counter.cpp)

```
1    #include "counter.h"
2
3    // Initialization of a static variable
4    int counter :: m_count = 0;
5
6    counter :: counter ()
7    {
8      m_count ++;
9    }
10
11   counter ::~ counter ()
```

```
12   {
13      m_count --;
14   }
15
16   /* static */ int counter :: getCount ()
17   {
18   return m_count;
19   }
```

Within the constructor, the static variable *m_count* is increased by the value 1. Since the variable is the same for all objects, it is counted how many objects of the class were created. So that this value is still correct when objects of the class are deleted, it is necessary to decrease the value of the variable *m_count* by 1 again within the destructor.

The function *getCount* was declared as a static function. As with *inline* and *virtual, static* is not repeated again in the function definition. Within the function, function parameters and static member variables and functions can be accessed. In this example, only the value of the variable *m_count* is to be returned.

In the main function of Listing 10.28, the functions of the class are now tested.

Listing 10.28 Test Program for the *counter* Class (project.cpp)

```
1    #include <iostream >
2    #include "counter.h"
3
4    using namespace std;
5
6    // Main function
7    int main ()
8    {
9      cout << counter :: getCount () << endl;
10
11     counter c1;
12     counter c2;
13     counter c3;
14
15     cout << counter :: getCount () << endl;
16
17     for (int i = 0; i < 5; i++)
18     {
19        counter c;
20        cout << counter :: getCount () << endl;
21     }
22
23     system (" pause ");
24   }
```

As can be clearly seen, the *getCount* function is called before the first object of the class has been created. The output corresponds to the value 0, since the static variable has only been initialized so far. Following this output, three objects of the class *counter* are created and again the return value of the function *getCount* is printed. The result is now 3, as expected, since the variable value was increased by 1 with each constructor call (Listings 10.29 and 10.30).

Within the following loop, an object of the class *counter* is created at each loop pass and the function *getCount* is called to print its return value. However, no ascending numbers 4, 5, ... are printed, but always the number 4. Since the object is defined within the loop body, the object only exists for exactly one loop pass. Thus, an object of the class is created with each run, but it is also deleted again immediately. Each time the constructor is called, the variable value is increased by 1, but at the end of the loop, the value is decreased by 1 when the destructor is called. The total number of currently existing objects does not change.

This successfully tests all the functions of the class. The complete output of the program is:

```
0
3
4
4
4
4
4
```

Exercises

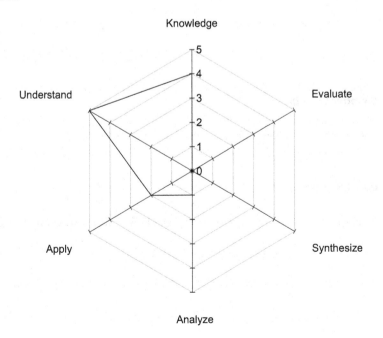

Network diagram for the self-assessment of this chapter

10.1   Visibility Levels

Name the different visibility levels for classes in *C++* and what they mean!

10.2   Operators

What is the meaning of the term operator in a class in *C++*?

10.3   *Include Guards*

What is an *include guard* and what is it designed to prevent?

10.4   Abstract Classes

State the meaning of the term "abstract class"!

10.5   *Member Variables*

Explain how *member variables* differ from other variables in a class!

10.6   Constructors

Explain the purpose of the constructor of a class and how it differs from other functions!

10.7   Classes and Structures

Identify differences between classes and structures in *C++*!

10.8   Polymorphism

Explain the concept of polymorphism!

10.9   Polymorphism

Explain the function of the keyword *static* for classes!

10.10   The Class *point2D*

Write your own class, which should be named *point2D*. The class should have two member variables m_x and, m_y both of which should be of type *double*. The visibility level of the member variables should prevent the variables from being accessed from outside the class.

Also develop three constructors for the class:

- The first constructor should do without parameters and initialize the two variables with the value 0.0 each.
- The second constructor should be able to fill the member variables with meaningful values during initialization. Therefore, the constructor should receive two variables x and y and copy the values into the member variables.
- The third constructor is to be a copy constructor.

Two points are to be added with the help of an operator +. The result should be a new point whose coordinates correspond to the sum of the two point coordinates.

The operator * is to calculate the scalar product by $s = \vec{a} \cdot \vec{b} = a_x {}^* b_x + a_y {}^* b_y$. The result is therefore a variable of type *double*.

Also, write an operator that you can use to print the contents of the class with *cout*. The output should be in the form: *point2D(x, y)*.

Try to use and test the different functions in your main program!

10.11   The *Circle* Class

In this task, another class is to be developed, which is to be named *circle*. The class is to have two member variables. The first variable *m_center*, which shall be of type *point2D* from Exercise 10.10, shall store the center of the circle. The second variable *m _ r* is to store the radius as a *double*.

Three constructors are to be implemented:

- The first constructor does not need any parameters and shall set the center to the coordinates (0.0; 0.0). The radius should have the value 1.0.
- With the second constructor an initialization should be possible, therefore three variables of the type *double* should be passed. Thus the *x*, *y* and *r* values are to be passed.
- The third constructor is also to be used for initialization. The passing parameters shall be the center *point*, as *point2D*, and the radius, as *double*.

The *area* function is to calculate the area of the circle using the formula $A = \pi \cdot r^2$.

With the function *perimeter* the circumference of the circle is to be determined by the formula $U = 2 \cdot \pi \cdot r$.

Also, write an operator that you can use to print the contents of the class with *cout*. The output should be in the form: *circle(point2D(x, y), r)*.

Try to use and test the different functions in your main program!

10.12   Program Analysis

Analyze the following program. To do this, try to find out what the individual program lines do in terms of content and deduce the task of the program.

Commands have been used that you don't know yet. Try to research them!

Do not type the program, but try to understand what is happening without assistance!

Riddle.h

```
1   // Include -Guard
2   #include <string >
3
4   using namespace std;
```

```
 5
 6    class Riddle
 7    {
 8    public:
 9      Riddle(string data );
10
11      friend ostream& operator <<( ostream &out , Riddle r);
12    protected:
13      string m_data;
14    };
15
16    ostream& operator <<( ostream &out , Riddle r);
```

Riddle.cpp

```
 1    #include "Riddle.h" #include <iostream >
 2
 3    Riddle :: Riddle(string data)
 4    {
 5        char k;
 6
 7        for (int i = 0; i < data.length (); i++)
 8      {
 9        k = data[i];
10
11        if (k >= 97 && k <= 122)
12          k = 65 + (k - 94) % 26;
13        else
14          if (k >= 65 && k <= 90)
15            k = 65 + (k - 62) % 26;
16
17      m_data += k;
18      }
19    }
20
21    ostream& operator <<(ostream &out , Riddle r)
22    {
23    char k;
24
25    for (int i = 0; i < r.m_data.length (); i++)
```

```
26      {
27        k = r.m_data[i];
28
29        if (k >= 65 && k <= 90)
30          k = 65 + (k - 42) % 26;
31
32          out << k;
33      }
34
35    return out;
36    }
```

Pointer

11

Short and Sweet

- The memory of a program is roughly divided into four areas
  - Program code
  - Global variables
  - The *stack* for functions and local variables
  - The *heap* for dynamically created memory areas
- Each memory location has an address
- Pointers make it possible to remember certain addresses
- **Possible sources of error:**
  Errors can easily occur when working with pointers.
- **Possible sources of error:**
  The errors caused by pointers are very difficult to find in most cases.

In all previous chapters, variables were created inside or outside functions or as part of classes. Most of these variables are stored in a specific area of main memory called the *stack*. The *stack* is subject to a number of constraints that allow it to be accessed quickly. First, the *stack* is limited in size and second, access always follows the so-called *Last In, First Out* (LIFO) principle. This means that the data that was last stored on the *stack* is always on top and is the first to leave the *stack*. The *stack* is available in its fixed size for the entire runtime of the program.

For each function call, data is stored on the *stack* for the respective function. This data includes the position to which the program must return when the function is terminated, the function parameters and the local variables of the function. When a function is terminated, all of this data is removed from the *stack* again. Since only one function can be

B. Tolg, *Computer science to the Point*,
https://doi.org/10.1007/978-3-658-38443-2_11

active at a time, the data of the current function is always on top of the *stack*. Thus, the existence of the local variables and parameters is linked to the execution time of the respective function.

Global variables, which are created outside of functions, must exist over the entire runtime of the program and are stored in their own memory area. The same applies to the program code, which is also stored in its own area of main memory.

However, very many applications require variables whose lifetime is not coupled to functions and whose size far exceeds that of the *stack*. In addition, these variables should not exist during the entire runtime of the program, but only when they are needed. An example of this is an application that is supposed to load and process images. The size of an image with the resolution 1920 · 1080 pixels is about 8 MB uncompressed at 4 bytes per pixel. The usual size for the *stack* is 1 MB. Thus it would not be possible to store such an image on the *stack*. In addition, applications of this type should be able to manage several images in the vast majority of cases, which can be loaded and closed again during the program runtime.

To enable such applications, memory can be requested dynamically, which is created on the so-called *heap*. Memory that is requested in this way is allocated to the program until the memory is explicitly released again by the program. Since in this case the program itself must remember where the requested memory is located, so-called *pointers* are required. Working with pointers opens many new possibilities, but also many new sources of errors, which are very difficult to find in programs. For this reason it is necessary to develop a basic understanding of working with pointers and memory.

First of all, in the memory of a computer, each byte is numbered consecutively. In principle, this numbering is comparable to the house numbers in a street, so that the term *address* is also used here. Each variable that is created in the program must be stored at an address and occupies a certain amount of memory there, depending on the variable type. Table 5.1 in Chap. 5 shows the memory consumption for each variable type. If, for example, a 4-byte variable such as an *int* is stored at a certain address, the next 3 bytes are also occupied by this variable and can no longer be used for other data (see Fig. 11.1).

The addresses of the individual bytes have been indicated by $N, N + 1, \ldots$ in this figure. In fact the hexadecimal number system is often used for the representation of adresses. Even if this representation is a hurdle at the beginning, it increases the readability of such addresses later on. Since each byte consists of 8 bits, a byte can always be represented by two hexadecimal digits.

Each variable created in the example programs is therefore located at a specific address within the memory. In *C++*, the address of a variable can always be accessed using the

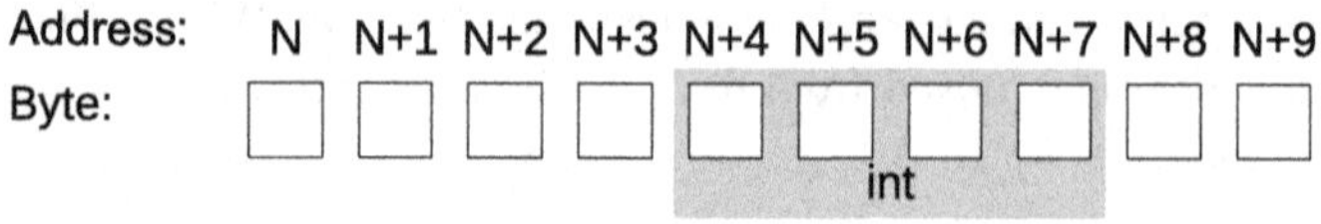

Fig. 11.1 Schematic representation of a variable in the main memory

ampersand symbol. In the main function of Listing 11.1, a variable of type *int* is created with the name *data* and initialized with the value 0.

Listing 11.1 Output of a Memory Address

```
1    #include <iostream >
2
3    using namespace std;
4
5    // Main function
6    int main ()
7    {
8        // Variable definition and initialization
9        int data = 0;
10
11       // Output of the memory address
12       // of the variable data
13       cout << &data << endl;
14
15       return 0;
16   }
```

The statement *cout* < < & *data* < < *endl*; is used to print the address where the data of the variable *data* is located on the console. The output of the program is.

```
0136FA74
```

Since the output is hexadecimal, the variable was therefore stored at the decimally represented memory address 20,380,276. The address is located in the *stack* of the program. If the memory contents were printed with the associated addresses, the result shown in Table 11.1 would be obtained.

Table 11.1 Memory allocation by Listing 11.1

Area	Address (hex)	Address (dec)	Data (hex)	Data (dec)
Stack	0136FA79	20,380,281	?	?
	0136FA78	20,380,280	?	?
	0136FA77	20,380,279	00	0
	0136FA76	20,380,278	00	0
	0136FA75	20,380,277	00	0
	0136FA74	20,380,276	00	0
	0136FA73	20,380,275	?	?
	0136FA72	20,380,274	?	?

In the surrounding memory areas there are any values that are not known. But at the 4 bytes, which are occupied by the *int* variable, there must be the numerical value 0, because all 4 bytes of the variable must represent this number.

To claim a memory area on the *heap*, the *new* instruction followed by a variable type must be used. This instruction attempts to allocate memory on the *heap* and returns the starting address of the memory area if successful. This address must now be saved in a suitable variable, a pointer, so that the memory can be accessed. To declare a pointer, a normal variable of the type to be at the saved address is first created. The only difference is that a * is inserted between the type and the name of the variable. It does not matter whether the * is inserted directly at the variable type, the name, or even separated by spaces from both.

The asterisk causes a pointer variable to be created whose memory consumption no longer has anything to do with the named type, but only with the size of a memory address. In a 32-bit system, an address consists of 4 bytes; in a 64-bit system, an address consists of 8 bytes. A pointer of the type *char** therefore requires just as much memory as a pointer variable of the type *int**, although the data types *char* and *int* differ in terms of memory consumption.

In Listing 11.2, a pointer variable of type *int** is created with the name *data* and initialized with the result of the *new* statement. The *new* statement creates a four-byte area of memory on the *heap* by adding *int(0)* and initializes it with the value 0. If the *new* statement is able to reserve the memory successfully, it returns the start address of the area as a return value.

Listing 11.2 Creating a Variable on the *Heap*

```
1    #include <iostream >
2
3    using namespace std;
4
5    // Main function
6    int main ()
7    {
8      // Variable definition and initialization
9      int* data = new int (0);
10
11     // Output of the memory address
12     // of the variable data
13     cout << &data << endl;
14
15     // Output of the memory address
16     // that was saved in data
17     cout << data << endl;
18
```

```
19      // Output of the value
20      // that was stored on the heap
21      cout << *data << endl;
22
23      // Release of the heap memory
24      delete data;
25      data = 0;
26
27      return 0;
28   }
```

Next, three outputs follow through the program. The pointer variable *data* is a normal variable that is stored on the *stack*. The address of this variable is printed first by the instruction *cout* < < & *data* < < *endl*;. The variable *data* stores an address that is on the *heap* and was generated by the *new* statement. This content of the variable *data* is printed by the statement *cout* < < *data* < < *endl*;. The last thing to be printed is the content of the memory address on the *heap*, i.e. 0. To achieve this, the name of the pointer variable is preceded by *. The variable is *dereferenced*. This name describes an indirect access. First, the contents of the variable *data* are interpreted as an address, then the contents of this address are interpreted as *int* (since *data* is a pointer of type *int**) and returned. The statement *cout* < < * *data* < < *endl*; prints the result to the console.

Finally, the memory area on the *heap* must be manually freed when it is no longer needed. This is done with the instruction *delete data*;. However, this statement only releases the memory area at the corresponding address. The now invalid address is still stored in the pointer variable *data*. Accessing this address again may cause an error, but under certain circumstances it may not. If the memory area is used by the program elsewhere in the meantime, another data set may be corrupted. This can lead to errors that only cause effects much later. For this reason, the best approach is to set a pointer variable to the value 0 immediately after the memory area has been released. This address is always invalid for access and will cause a program crash should an attempt be made to access it. This sounds like a problem because a program crash is obviously undesirable. In fact, however, an error is desirable in this case, because it is located exactly at the point where the program crashes. Thus, the error can be quickly detected and corrected.

This approach is also useful when pointers are not used immediately. A golden rule should be: A pointer either has a valid value, or it contains the value 0.

The program generates the following output:

```
0115F7C0
03232E10
0
```

Figure 11.2 shows the memory allocation of Listing 11.2 schematically. Table 11.2 shows the contents of the individual memory locations in both hexadecimal and decimal

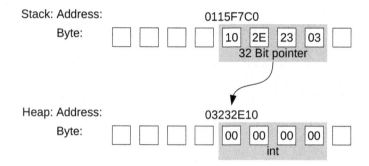

Fig. 11.2 Schematic representation of the memory allocation by Listing 11.2

Table 11.2 Memory allocation by Listing 11.2

Area	Address (hex)	Address (dec)	Data (hex)	Data (dec)
Heap	03232E15	52,637,205	?	?
	03232E14	52,637,204	?	?
	03232E13	52,637,203	00	0
	03232E12	52,637,202	00	0
	03232E11	52,637,201	00	0
	03232E10	52,637,200	00	0
	03232E0F	52,637,199	?	?
	03232E0E	52,637,198	?	?
	...	...	...	...
Stack	0115F7C5	18,216,901	?	?
	0115F7C4	18,216,900	?	?
	0115F7C3	18,216,899	03	3
	0115F7C2	18,216,898	23	35
	0115F7C1	18,216,897	2E	46
	0115F7C0	18,216,896	10	16
	0115F7BF	18,216,895	?	?
	0115F7BE	18,216,894	?	?

notation. The content of the variable *data* is located on the *stack* from address 0115F7C0. If the content of the memory is read from the back to the front, the result is the address 03 23 2E 10, which is located on the *heap* and in which the value 0 was stored.

Here another advantage of the hexadecimal notation becomes apparent. In hexadecimal, the values of the individual bytes can be directly appended to each other, so that they result in the target address. The values of the decimal notation do not allow this direct translation into a decimal address.

11.1   Type Conversion

The type of a pointer has a different meaning than the type of a "normal" variable. In the case of the "normal" variable, the variable type determines the size of the memory consumption and the way in which the content of the variable is interpreted. A *double* is a real number that occupies 8 bytes of memory according to Table 5.1.

With a pointer variable, the size of the memory space is determined exclusively by whether the program is created for 32-bit systems or for 64-bit systems. The memory consumption is 4 bytes in the first case and 8 bytes in the second. The type of the pointer variable is only used for how the content of the stored target address is to be interpreted. If a pointer has the type *int** then, starting from the stored address, the next four bytes are combined and interpreted as an integer.

For this reason it is possible to create a pointer of type *void**. With a "normal" variable, the *compiler* could not decide how much memory to provide for the variable. Therefore, there can be no variables of type *void*. But with a *void* pointer,* the size is already fixed. However, it would not be possible for the *compiler* to interpret the contents of this pointer. The use of *void** pointers is more widespread than it might seem at first glance. They are always used when the content of a pointer does not need to be edited, or there can be different types of pointers that are to be processed in a function, for example.

To process the contents of a *void** pointer, the variable type must be changed by a type conversion so that the *compiler* can interpret the contents. In Listing 11.3, a pointer of type *void** is created and initialized using a *new* statement. The *new* statement, of course, requires a variable type to create memory on the *heap*. In this example, the *new* statement creates and returns an address of type *int**. Since the *void** pointer makes no assumption about the contents of the memory address, value assignment in this direction is allowed.

Listing 11.3 Type Conversion of a Pointer Variable

```
1    #include <iostream >
2
3    using namespace std;
4
5    int main ()
6    {
7        // The value assignment from int* to void* works
8        void* pointer = new int (25);
9
10       // The value assignment from void* to int* works
11       // only if the type is explicitly converted.
12       int* pointer2 = (int*) pointer;
13
14       // The direct output causes an error
15       // cout << *pointer << endl;
```

```
16
17      // If the type of the pointer is changed ,
18      // so the output is not a problem.
19      cout << *(int*) pointer << endl;
20
21      delete pointer;
22      pointer = 0;
23      pointer2 = 0;
24
25      return 0;
26   }
```

Next, a variable *pointer2* of type *int** is to be created by *int*po int er2* = (*int**)*po int er*; and initialized with the content of the variable *pointer*. Since *pointer2* makes the assumption that there is a value of type *int* at the stored destination address, but the variable *pointer* makes no such assumption, the variable *pointer* must be explicitly converted to type *int**. Otherwise, the *compiler* will generate an error message.

Dereferencing the variable *pointer*, as in the commented out line *cout < < * zeiger< < endl;*, is also not allowed because of the uninterpretable type *void**.

However, if the pointer is converted to *int** type as in the line *cout < < * (int*)po int e r< < endl;*, the result can be interpreted and printed. It is noteworthy that first the type of the pointer is converted and only then it is dereferenced. As expected, the program generates the output 25.

Finally, the created memory must be released again. Since the variables *pointer* and *pointer2* refer to the same address, the *delete* statement may of course only be applied to one of the two variables. Afterwards, however, both variables should receive the value 0 in order to prevent incorrect accesses to the released memory.

11.2   *const*

Since pointers, just like references, allow direct access to memory, it is also necessary here to have the option of preventing write access. As with other examples, the *const* keyword can also be used for pointers.

If a simple pointer variable is declared, there are three different positions where the keyword *const* may be placed. However, two of these positions lead to the same result. In Listing 11.4, the different variants of constant pointer variables were created.

Listing 11.4 Various Constant Pointer Variables

```
1    #include <iostream >
2
3    using namespace std;
```

```
4
5    int main ()
6    {
7        // Value constant , address variable
8        const int * pointer1 = new int (0);
9        // Value constant , address variable
10       int const * pointer2 = new int (0);
11       // Value variable , address constant
12       int* const pointer3 = new int (0);
13
14       // *pointer1 = 5; // Not OK
15       // *pointer2 = 5; // Not OK
16       *pointer3 = 5;
17
18       cout << *pointer1 << endl
19             << *pointer2 << endl
20             << *pointer3 << endl;
21
22       delete pointer1;
23       delete pointer2;
24       delete pointer3;
25
26       pointer1 = 0;
27       pointer2 = 0;
28       // pointer3 = 0;
29
30       return 0;
31   }
```

The first two lines *const int * pointer1 = new int(0);* and *int const * pointer2 = new int(0); do* exactly the same thing. In both cases, a pointer variable is created where it is not allowed to change the contents of the stored address. For this reason, the lines *po int er*1 = 5; and *po int er*2 = 5; are commented out. But it is allowed to change the address itself. It would therefore be possible to assign the address of another *int variable* to the variable, but its content would not be allowed to be changed either.

Both notations refer to the *int,* the second notation being the one that is easier to understand. Normally, the keyword *const* always refers to the element that is to the left of the keyword. Only for the first element, in this case the *int,* it is additionally allowed to drag the keyword in front of the element it refers to.

By the line *int* const pointer3 = new int(0);* a pointer variable is initialized, where the content of the memory address may be changed, but not the stored address. So it is easily possible to change the value on the *heap* from 0 to 5 by using the line *po int er*3 = 5;. Also in this notation, the keyword refers to the element directly to the left of it, in this case the star symbolizing the address.

The output of the program is.

```
0
0
5
```

because only the content of the memory area pointed to by *pointer3* was allowed to be changed.

After the memory areas have been released again by the *delete* statements, the stored addresses are to be set to 0 in order to prevent erroneous accesses. However, since the address stored in *pointer3* is constant, the line *po int er*3 = 0; cannot be executed.

Of course, it is also possible to combine both expressions to create a pointer where neither the address nor its contents may be changed. The corresponding line would be *int const * const pointer4 = new int(0);*.

11.3   Arrays

With the previous arrays it was not possible to let the size of the array be entered during the runtime of the program, because the *compiler* only accepted constant array sizes. However, especially with arrays it makes sense to adapt the size individually to the requirements. If, for example, a data set was created that was to store 1 MB of data, the program would crash immediately, since no further memory would be available on the *stack*. Increasing the size of the *stack* would also be an insufficient solution, because on the one hand this memory is permanently occupied by the program and on the other hand the same problem would then simply occur a little later. In addition, memory of this size is often only needed when the program has to load and process a certain data set (perhaps an image). When that record is no longer needed, the memory can be freed. In Listing 11.5, an array of dynamic size is created. To do this, an auxiliary variable *n* is first created to store the size of the array. Unlike Listing 8.1, the variable *n* no longer needs to be declared as *const*. Next, the program is to read in a value for the array size *n*.

Listing 11.5 Creating a Dynamic Array

```
1    #include "stdafx.h"
2    #include <iostream >
3
4    using namespace std;
5
6    int main ()
7    {
8       // Variable definition and initialization
9       // for the array size
```

```
10      unsigned int n = 0;
11
12      // Entering the array size
13      cout << "Please enter the size "
14          << "of the record:" << endl;
15      cin >> n;
16
17      // Initialization of a dynamic array
18      int* data = new int[n];
19
20      // Initialization of all array values with 0
21      for (int i = 0; i < n; i++)
22      {
23        data[i] = 0;
24      }
25
26      // Release Memory
27      delete [] data;
28      data = 0;
29
30      return 0;
31    }
```

Now a pointer variable *data* of type *int** is created and initialized using the *new* statement. The use of square brackets after the variable type makes it clear that an array is to be created on the *heap*. The variable *n* specifies the number of elements and the variable type *int* specifies the memory consumption per element. Thus, the statement *new int[n]* attempts to allocate *n*4Bytes* on the *heap*. After initialization, arrays on the *stack* and on the *heap* behave exactly the same again.

To clarify this, all elements of the array *data* are initialized with the value 0 in the program by the following *for-loop*. To do this, all indices 0, …, *n* − 1 are passed through once. As with all arrays, access is made by combining the array name and the index in square brackets. In this example, by *data[i]*.

However, since the array was created on the *heap*, the memory must be released manually, as with all pointer variables. This is done by the instruction *delete[]*, which makes it clear that not only the specified address, but also an array is to be released again.

Figure 11.3 shows the memory allocation of the program directly after the initialization of the array. A 4-byte address is stored on the *stack* in a 32-bit system. In the program, this is the pointer variable *data*. The content of this variable points to an address on the *heap*. There, because of the chosen variable type *int*, groups of 4 bytes are always interpreted as an integer. If the value 6 is chosen for the size of the array *n*, the situation shown results. For reasons of clarity, only the initial addresses of the respective integer values have been drawn in the figure.

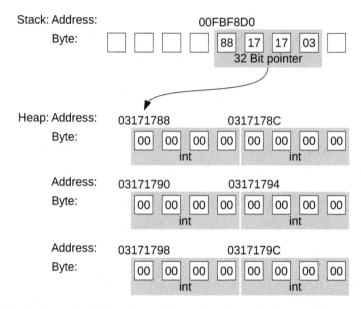

Fig. 11.3 Schematic representation of the memory allocation by Listing 11.5 for $n = 6$

11.4 Pointer Arithmetic

In the C++ language, all pointers behave very similarly to arrays. It is even possible to use the square brackets even if a pointer was not created as an array. This is because when square brackets are written after a pointer, the target address being searched for is calculated using a simple formula that can be applied in any case. If a program contains the line.

```
pointer[n] = 15;
```

then the formula with which the destination address can be calculated is always

$$(Destination) = (addressSize) + n \cdot (of\ the\ data\ type) \qquad (11.1)$$

With this notation, access is always indirect. This means that the content of the target address is changed, just as with an array.

This type of memory access is very dangerous, since it is absolutely necessary to know exactly which positions can be accessed safely. It would be very easy to access behind a reserved memory area and thus generate arbitrarily complex errors. However, this form of memory access also offers great advantages, especially if the type of the pointer is changed. The following Listing 11.6 shows an example of such an access.

Listing 11.6 Output of a Memory Area Byte by Byte

```
1    #include <iostream >
2
3    using namespace std;
4
5    int main ()
6    {
7        // Initialize an int pointer
8        int *data = new int (4223);
9
10       // Initialize an unsigned char pointer
11       // with the address of data
12       unsigned char* bytes = (unsigned char *) data;
13
14       // Output of stack position and contents
15       cout << "Stack: " << &data << " : " << data << endl;
16
17       // Output of heap positions and contents
18       // Byte by byte
19       for (int i = 0; i < 4; i++)
20       {
21           cout << "Heap: " << (int *)& bytes[3-i]
22                << " : " << (int)bytes[3-i] << endl;
23       }
24
25       // Release Memory
26       delete data;
27       data = 0;
28
29       return 0;
30   }
```

First, a pointer of type *int** is created by the line *int *data = new int(4223);* and initialized using the *new* statement. On the *heap*, the memory space for a 4-byte variable of type *int* is created and the reserved address is stored in *data*. In the line *unsigned char* bytes = (unsigned char*)data;* another variable with the name *bytes* is created and initialized with the same address to which *data* already points. For this to work, a type conversion from *int** to *unsigned char** must be performed on *data*.

Both variables now differ when accessing the memory contents using the square brackets, since formula 11.1 depends on the size of the data type. Since *data* is of type *int**, the address would be increased by the value 4 by *data[1]*. In contrast, *bytes* is of type *unsigned char**, so *bytes[1]* would only increase the address by the value 1. This makes it possible to look at the bytes that make up the integer value in memory individually.

The line *cout < <'Stack:" < < & data < <':' < < data < < endl*; first prints the address at which the variable *data* is stored on the *stack* and then, followed by a colon, the value stored at this address. The latter, of course, corresponds exactly to the address on the *heap* at which the storage location for the integer value was created.

In the following *for loop*, the count variable *i* now runs through the values 0, 1, 2, 3 in order to be able to output the individual bytes of the integer. The statement *(int*)&bytes[3 − i]* achieves the following: *bytes[3 − i]* accesses the address that was increased by 3 − *i* bytes. This serves only the cosmetic purpose that the values are to be printed in descending order from top to bottom, thus starting at the largest address. The statement *&bytes[3 − i]* now does not print the content of the corresponding address, but the address itself. However, since it is a variable of type *unsigned char**, *cout* tries to print it using text characters. To prevent this, *(int*)&bytes[3 − i]* performs a type conversion to an *int** so that a hexadecimal address is printed.

Almost the same meaning has the instruction *(int)bytes[3 − i]*. The only difference is that the value is to be printed here, not the address. The type conversion to the type *int* is carried out accordingly.

Finally, the created memory space on the *heap* is released again and the stored address is overwritten with 0.

The output of the program looks like this:

```
Stack:  00AFF7EC  :  00DBCEA0
Heap:   00DBCEA3  :  0
Heap:   00DBCEA2  :  0
Heap:   00DBCEA1  :  16
Heap:   00DBCEA0  :  127
```

On the *stack* at address 00AFF7EC the address 00DBCEA0 is stored, which is located on the *heap*. A variable of type *int* was created at this address, which is stored at addresses 00DBCEA0 to 00DBCEA3. If the contents of these memory addresses (i.e. 0 0 16,127) are represented hexadecimally, the result is the number 00 00 10 7F and this corresponds exactly to the expected number 4223.

Pointers can also be modified using the increment and decrement operators ++ and −−. In addition, values can be added and subtracted. With all these operators, however, it is not the specified value *n* that is added or subtracted, but *n* times the variable type of the pointer. This sounds complicated at first, but in practical use it is quite simple. When writing a program, it is not necessary to pay attention to the size of the variable type to access the next element. In any case a 1 is added. Internally, the pointer is then increased by the required number of bytes.

So the *for loop* in Listing 11.6 could also be replaced with the following source code:

```
// Output of heap positions and content
// Byte by Byte
bytes += 3;

for (int i = 0; i < 4; i++)
{
  cout << "Heap: " << (int*)bytes << " : "
       << (int)*bytes << endl;
  bytes--;
}
```

The output of the program after this change would be identical to that of program 11.6. However, the second variant is faster, because the access can be made directly to the address stored in the variable *bytes*, whereas in the first variant the target address has to be calculated twice by formula 11.1.

11.5   Advanced: Multidimensional Arrays

Multidimensional arrays are even a bit more challenging if they are to be created on the *heap*. This is due to the fact that there are several variants that can be used to create a multidimensional array in memory. Each of these variants has very specific advantages and disadvantages that affect the properties of speed, flexibility, and comprehensibility.

It is therefore important to understand exactly how *C++* works with the different variants in order to make the right choice for your own problem.

11.5.1   Variant 1: Pointer to Arrays

One way to create *N-dimensional* arrays in *C++* is to create a pointer to an $N - 1$-dimensional array. The advantage of this variant is that the memory is reserved contiguously. Thus, memory accesses always occur in the same region of memory, which speeds up access. However, only the size of the first dimension is freely and dynamically selectable. The sizes of the other $N - 1$dimensions must be as constant as if they were reserved on the *stack*. This is due to the fact that the $N - 1$ dimensions of the array in this variant are implemented using the variable type of the pointer. This must have a constant size so that formula 11.1 can be applied.

In addition, the notation for reserving such an array, while consistent within the language, is not immediately obvious. In Listing 11.7, a two-dimensional array is created on the *heap* by creating a pointer of variable type *int[1024]**.

Listing 11.7 Creating a Multidimensional Dynamic Array

```
1    int main ()
2    {
3      // Variable definition and initialization
4      // for the array size
5      unsigned int sizeY = 1024;
6      const unsigned int sizeX = 1024;
7
8      // Initialization of a dynamic array
9      int (* data )[ sizeX] = new int[sizeY ][ sizeX ];
10
11     // Initialization of all elements
12     for (int i = 0; i < sizeY; i++)
13     {
14       for (int j = 0; j < sizeX; j++)
15       {
16         data[i][j] = 0;
17       }
18     }
19
20     // Release Memory
21     delete [] data;
22     data = 0;
23
24     return 0;
25   }
```

First, the dimensions of the array are set to the value 1024 by the variable *sizeY* and the constant *sizeX*. The constant sizeX serves as the constant size of the pointer type.

The initialization of the array is performed by line 8 *int (*data)[sizeX] = new int[sizeY] [sizeX];*. Here, three conventions of the *C++* language clash, making the resulting notation difficult to read. The first convention states that the type of the pointer must always precede the star symbol. The second convention states that the dimension of an array must always come after the name of the array, and the third convention states that the asterisk belongs to the type of the pointer, not its name. These three conventions conflict here. The solution *int (*data)[sizeX]* is the resolution of the conflict. The variable type is specified by *int* at the beginning and *[sizeX]* at the end. This satisfies the conventions for the variable type and the position of the array dimension. Now it must be made clear that a pointer is to be created. This is done by defining the name of the variable in round brackets preceded by an asterisk. The result is a pointer variable with the name *data* and the variable type *int[sizeX]**.

After the unusual initialization of the array, the pointer to the array can be used as it is already known from two-dimensional array from the *stack*. Within a double *for-loop* starting at line 11, all elements of the array are assigned the value 0. The line *data[i]* *[j] = 0;* can be interpreted as follows: *data[i]* returns a pointer to an array of type *int* and size *sizeX* = 1024. Because of the variable type *int[1024]*, the address stored in *data* is incremented by $i{*}4096$ bytes according to formula 11.1 and thus points to the $i + 1$-th array. Then *[j]* accesses the element that is located $j{*}4$ bytes after this address, and this is exactly the element at position *[i][j]*.

Finally, the memory is released again and the pointer is set to address 0.

Figure 11.4 shows how the memory is occupied by this variant.

The pointer on the *stack* is located at address 00F3FBF0 and points to address 035C4040, which is located on the *heap*. Starting at this address, a contiguous block of memory has been reserved, consisting of 1024 blocks along the vertical axis, all of which have the variable type *int[1024]* and thus consist of 4096 bytes. As a result, the addresses at the beginning of the line always differ by the value 4096.

This variant offers the advantage that the memory is reserved contiguously and, apart from the unusual initialization, can be addressed in exactly the same way as other multidimensional arrays. However, only the size of the first dimension is freely selectable. All other dimensions must have a constant size. Care must also be taken when initializing *int (*data)[sizeX] = new int[sizeY][sizeX];* that the array dimensions on the left side of the equal sign have the same size and order as in the memory request on the right side of the equal sign. For example, expansion to three dimensions would be achieved by *int (*data) [sizeY][sizeX] = new int[sizeZ][sizeY][sizeX];* with variable *sizeZ* and constant *sizeX* and *sizeY*.

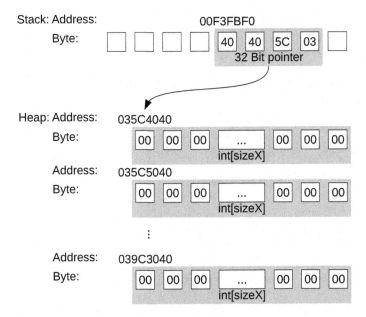

Fig. 11.4 Schematic representation of the memory allocation by Listing 11.7

11.5.2   Variant 2: Pointer to Pointer

The second variant can be seen as a further development of the one-dimensional arrays on the *heap*. Up to now, a pointer was created on the *stack* that pointed to an address on the *heap* at which memory space was reserved for an array. If this array were to consist of pointers that point to arrays on the *heap*, it would be possible to create a two-dimensional array whose extent could even be selected individually for each line.

The idea can also be extended to arbitrary dimensions by pointers pointing to pointers pointing to pointers, and so on. However, the disadvantage of this variant is that the reserved memory is not contiguous. For this reason, accessing each element may be slowed down by accesses at different positions in the memory. In addition, with each new dimension a new level of pointers is added, making it difficult to understand the resulting data structure, especially at the beginning.

In Listing 11.8, a two-dimensional array is created by applying variant 2 described in this chapter. To do this, two auxiliary variables *sizeX* and *sizeY* are again first created and initialized with the value 1024. None of the array dimensions need be constant, and in fact it would even be possible to use a different value for *sizeX* in each line. But the example should not become unnecessarily complicated.

Listing 11.8 Creating a Multidimensional Dynamic Array (Advanced Variant)

```
1    int main ()
2    {
3      // Variable definition and initialization
4      // for the array size
5      unsigned int sizeY = 1024;
6      unsigned int sizeX = 1024;
7
8      // Initialization of a dynamic array
9      int** data = new int*[ sizeY ];
10
11     // Initialization of all arrays
12     for (int i = 0; i < sizeY; i++)
13     {
14       data[i] = new int[sizeX ];
15     }
16
17     // Initialization of all elements
18     for (int i = 0; i < sizeY; i++)
19     {
20       for (int j = 0; j < sizeX; j++)
21       {
22         data[i][j] = 0;
```

```
23          }
24      }
25
26      // Release of all arrays
27      for (int i = 0; i < sizeY; i++)
28      {
29        delete [] data[i];
30        data[i] = 0;
31      }
32
33      // Release Memory
34      delete [] data;
35      data = 0;
36
37      return 0;
38  }
```

The eighth line *int** data = new int*[sizeY];* creates a new pointer variable *data* of type *int***. This notation seems strange at first glance, but it consistently continues what has already been described for the previous pointer variables. The second asterisk specifies that it is a pointer variable, preceded by the type that is expected at this address. And in this case, this is again a pointer, which again points to a value of type *int*. So a value of the type *int**.

This notation is also consistently continued with the *new* statement. Here, memory is requested on the *heap* for an array of type *int**. This is an array of pointers that can point to values of type *int*. However, this is only the first step of the initialization, because behind the now created pointers there is of course still no memory in which values can be stored.

This does not happen until the following *for loop* in line 11. Here, all pointers of the array are now initialized by *data[i] = new int[sizeX];*. This again requests memory on the *heap* for an array of type *int* with *sizeX* elements. The returned address is stored in the *i-th* pointer of *data*. Through this two-step process, it is also obvious why the reserved memory is not necessarily contiguous. Memory is reserved by *sizeY* + 1 individual requests. The requested memory blocks are contiguous, but they are always reserved where there is just enough space, so that in the worst case the array is spread over *sizeY* + 1 disjoint areas.

Figure 11.5 shows the memory allocation generated by Listing 11.8.

The variable *data* was created on the stack at address 00 *CF F9* 64 and the address of the memory created on the *heap* 03 06 *F0* 88 was stored in it.

At this address there are now as many pointers in the *heap* as were specified by *sizeY*. These pointers are exceptionally displayed on top of each other in this illustration in order to have space for the next pointers. Each of these pointers points to its own memory address, where an array with elements of type *int* has been created. The size of the array corresponds exactly to the value of *sizeX*.

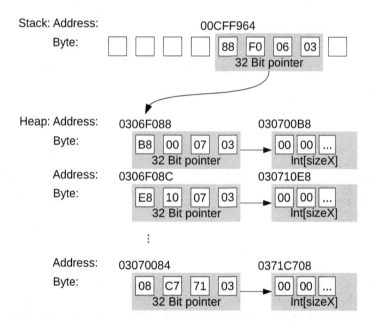

Fig. 11.5 Schematic representation of the memory allocation by Listing 11.8

For each arrow in the figure, a *new* statement was used, which could be located in a different memory area. If the number of arrows in this variant is compared with those from variants 1 or 3, it is immediately apparent that this solution is the least contiguous and thus potentially the slowest.

However, the access to the individual elements of the array is also carried out in this variant according to the already known scheme. Within the two loops in lines 17 and 19 the value 0 can be assigned to the elements of the two-dimensional array by *data[i] [j]* = 0;. In doing so, *data* returns a pointer to an array whose (*i* + 1) th element is accessed. The result is again a pointer to an array. Here, the (*j* + 1) -th element is accessed by assigning the value 0.

In this variant, the memory must also be released in two steps. First, starting at line 26, the memory of all lines must be released in a loop by *delete[] data[i];*. Here, too, the stored address should be replaced with the value 0 by *data[i]* = 0; in order to prevent errors.

Only then the memory from line 33 for the array of pointers can be released again by *delete[] data;* and protected against incorrect access by *data = 0;*. With a different order, the addresses of the lines would be lost and the memory could no longer be released.

This variant makes it possible to design very flexible array. At the same time, this approach is slow and a lot of pointers are used. This ensures that a certain sequence must be followed when creating and deleting the array. If this sequence is not adhered to, situations arise in which memory can no longer be released.

11.5.3   Variant 3: Virtual Dimensions

In addition to the variants already presented, there is a third possibility to create multidimensional arrays. While in the first variant the size of a dimension must be constant and in the second variant at the expense of the speed even every single line can have its own dimension, the third variant is flexible and fast in all dimensions. To achieve this, we first consider how a multidimensional array works in the first place. In memory all bytes are numbered in a row, so a very long wall-shelf is more suitable as a model, because it has only one extension. So a multidimensional array must somehow fit into this long wall shelf.

Figure 11.6 shows how the bytes can be arranged in a two-dimensional array. The fields with grey background represent the extension of the array in x and y direction. Thus the array is to have the size $N \cdot N$, using $N = 8$, with the indices i and j, *which* can assume the values $0, \ldots, 7 = N - 1$ respectively.

The fields with a white background represent the bytes that are located one after the other in memory and are simply numbered consecutively. It is easy to see that the number of the first byte in each line always corresponds to a multiple of $N = 8$. Even more precisely, the number of the first byte always corresponds exactly to $i \cdot N$. If the value of j is added, the exact number of each byte within the array can be determined depending on i and j.

If a two-dimensional array is to be mapped into a one-dimensional memory, this can be done line by line. The position of the byte within the linear memory is given by formula 11.2.

$$Byteposition \ i, j \quad i \ sizeX \ j \qquad (11.2)$$

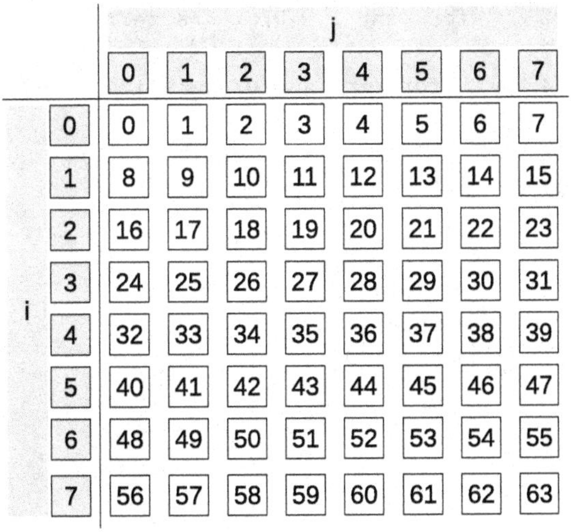

Fig. 11.6 Structure of a multidimensional array in a one-dimensional memory

This makes it possible to create a virtual two-dimensional array by creating a one-dimensional array of size $sizeY \cdot sizeX$. The position of a byte within a two-dimensional array can be calculated depending on i and j using formula 11.2. It is also possible to extend this solution approach to other dimensions. For example, a third dimension can be added by placing another array on top of the existing one. The formula would then be extended to formula 11.3.

$$Byteposition \ i,j,k \quad k \ sizeY \ sizeX \ i \ sizeY \ j \qquad (11.3)$$

The spatial notion of dimensions >3 is difficult, but the extension of the formula is easy to perform.

Listing 11.9 shows how this third variant can be implemented in *C++* for a two-dimensional array. First, two auxiliary variables *sizeX* and *sizeY* are initialized with the value 1024 in this program as well in order to define the dimensions of the array.

Listing 11.9 Creating a Virtual Multidimensional Dynamic Array (Programmer's Variant)

```
1    int main ()
2    {
3      // Variable definition and initialization
4      // for the array size
5      unsigned int sizeY = 1024;
6      unsigned int sizeX = 1024;
7
8      // Initialization of a dynamic array
9      int* data = new int[sizeY*sizeX ];
10
11     // Initialization of all elements
12     for (int i = 0; i < sizeY; i++)
13     {
14       for (int j = 0; j < sizeX; j++)
15       {
16         data[i * sizeX + j] = 0;
17       }
18     }
19
20     // Release Memory
21     delete [] data;
22     data = 0;
23
24     return 0;
25   }
```

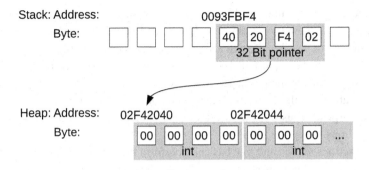

Fig. 11.7 Schematic representation of the memory allocation by Listing 11.9

Now, line 8 *int* data = new int[sizeY*sizeX];* creates a one-dimensional array with size *sizeY · sizeX* on the *heap. The* address of the array is stored in the pointer variable *data.*

In order to be able to initialize the elements of the array, it would of course be possible to create a single loop whose counting variable runs through all indices from 0 to *sizeY · sizeX* − 1. For this example program, however, two nested loops are to be used from line 11 onwards in order to clarify the access via two coordinates. As in the examples before, the variable *i* is used for the row and the variable *j* for the column. The elements of the array are accessed using formula 11.2, which converts the two-dimensional coordinates into the one-dimensional array position. The corresponding line in the program, in which all elements of the array are assigned the value 0, is *data[i * sizeX + j] = 0;*.

Figure 11.7 shows the memory allocation of Listing 11.9. In principle, the memory allocation is identical to that of Listing 11.5, except that in this example the number of elements was made dependent on the two dimensions of the virtual array. On the stack, the *heap* address of the reserved memory area 02 *F4* 20 40 is stored in the variable *data* at address 00 93 *FB F4*. This is followed by contiguous *sizeY · sizeX* values of type *int* whose addresses are always 4 bytes apart.

In this program, too, the reserved memory space is released again at the end by delete*[]* *data;* and the stored address is deleted by *data = 0;*.

Using this solution, it is very easy to create dynamic multidimensional arrays, which can also be addressed as one-dimensional arrays if it offers advantages. The only disadvantage of this solution is that the conversion of the two-dimensional coordinates has to be done manually. However, with the help of formula 11.2 this is no problem and an extension to higher dimensions is also easily possible.

11.6   Advanced: *malloc, realloc, free* and *memcpy*

For certain applications, the dynamic reservation of a specific memory area with flexible size is not yet sufficient to accomplish the task. If the size of the required memory can only be estimated when the program is started, it may be necessary to adjust the size of the reserved memory at runtime. In this case it is necessary to request new memory later. However, this new memory should complement and be related to the existing memory, which is not always possible. If such a situation arises, a completely new memory area of the desired size must be requested. The contents of the old memory area must be copied to the new area and the old memory space released.

In the *C* language, there are the functions *malloc, realloc, free* and *memcpy for* this application purpose. Thus *C++,* as a further development of *C,* can of course also handle these instructions.

The instructions *malloc* and *free* basically perform the same tasks as *new* and *delete* in C++. With *malloc* memory is reserved on the *heap* and with *free* this memory is released again. The two statements *new* and *delete,* however, can be overwritten as operators of a class. The functionality of *new* and *delete* is therefore not guaranteed. For this reason, you should never use *new* and *delete,* or *malloc* and *free,* together.

The function *realloc* makes it possible to increase or decrease an already reserved memory area. If a larger memory area is requested that cannot be accommodated at the previous position, a new area is reserved in the memory and the contents are copied from the old to the new position. The old memory area is released in this case. It is important to note that if no new memory can be reserved, a pointer to address 0, a *null pointer,* is returned. However, in this case, the old memory is not freed. For this case, the old address should be preserved in any case.

The *memcpy* function can copy an entire memory area from one address to another. Since the instruction copies a usually large memory area and does not interpret the contents of the memory, this instruction can be executed quickly.

Listing 11.10 creates a one-dimensional array using *malloc* and *free.* The program is to accept numbers from the console until the number 0 is entered. In this case, the input is to be aborted. To be able to store all numbers, a one-dimensional array is needed, but its required size is unknown. Of course, no memory should be wasted, reserving 1 GB of main memory just for this task would certainly be unreasonable. Nor would it be sufficient for a very patient person. The solution is an array whose size can be adjusted as needed.

Listing 11.10 Creating a Growing Dynamic Array with *malloc* and *realloc*

```
1    #include <iostream >
2
3    using namespace std;
4
5    int main ()
6    {
7      // Variable definition and initialization
8      unsigned int index = 0;
9      unsigned int value = 0;
10     unsigned int size = 5;
```

```
11
12        // Memory reservation with malloc
13        int* data = (int*) malloc(size * sizeof(int ));
14
15        do
16        {
17          // User input number unknown
18          cout << "Please enter a value: " << endl;
19          cout << "( end input with 0)" << endl;
20          cin >> value;
21
22          // When the end of the reserved area is reached
23          // a size adjustment must be made.
24          if (index >= size)
25          {
26            size += 5;
27            data = (int*) realloc(data , size * sizeof(int ));
28          }
29
30          // save the value
31          data[index] = input;
32
33          // new writing position
34          if (value != 0)
35          {
36            index ++;
37          }
38        } while (value > 0);
39
40        // Release Memory
41        free(data );
42        data = 0;
43
44        return 0;
45    }
```

First, some variables are created and initialized in the program. In the variable *index the* position is to be stored, at which the next number within the array must be stored. Initially, this is of course the position 0. At the same time, the variable of course also remembers how many values have already been stored. This would make sense if all values are to be printed later.

The value from the console is to be read into the variable value and the variable *size* is to store the actual size of the array. At the beginning of the program it is assumed that an average person enters a maximum of 5 numbers and then loses interest.

By line 13 *int* data = (int*)malloc(size * sizeof(int));* memory is now reserved on the *heap*. The function *malloc* takes as parameter the number of bytes to be reserved. In this case *size · sizeof(int)*. The function *sizeof* always returns the size of the passed data type in bytes, in this example 4. The return value of the function *malloc* is always a pointer of type *void*,* since in this case it is to be stored in a pointer of type *int*,* an explicit type conversion must take place.

Now, starting at line 15, a *do-while loop* follows, which continues to run as long as the value of the variable *value* is greater than 0. Within the loop, an action statement is first printed to the console and a value is read into the variable *value.*

Next, in line 24, *if (index size) >=* checks whether the current position to be written to is still within the previously reserved limits of the array. If this is not the case, the size of the array is increased by the value 5. The size of the reserved memory is then adjusted to the new size by *data = (int*)realloc(data, size * sizeof(int));.* The *realloc* function takes two parameters. The first parameter is the address whose memory area is to be changed. The second parameter is the new size of the memory area in bytes. Again, the return type is a *void** type pointer, so an explicit type conversion must be performed.

In order not to make the program unnecessarily complicated, it was omitted here to save the old address before the new memory area is reserved. In a "real" application this would have to be done of course.

After the *if statement*, it is ensured that the array is in any case large enough to store the new value. This is inserted into the array in line 31 by *data[index] = input;* at the position *index*. Finally, it is checked whether a value other than 0 was entered and, if necessary, the position of the *index* is increased by the value 1.

At the end of the program, the reserved memory is released again by *free(data);.* The function *free* takes the address of the memory area to be released as a parameter, but leaves the value of the pointer variable untouched. Therefore, the value of the address should also be deleted here by *data = 0;.*

Listing 11.11 provides a simple example of copying a memory area using *memcpy.* This is done by first using *malloc* to create two arrays on the *heap*, each of size *size = 100*.

Listing 11.11 Copying a Memory Area with *memcpy*

```
1    int main ()
2    {
3        // Variable definition and initialization.
4        unsigned int size = 100;
5
6        // Memory reservation with malloc
7        int* source = (int*) malloc(size * sizeof(int));
8        int* destination = (int*) malloc(size * sizeof(int));
9
10       // Initialization of the array
```

```
11      for (int i = 0; i < size; i++)
12      {
13        source[i] = 0;
14      }
15
16      // Copying the memory contents
17      memcpy(destination, source, size * sizeof(int));
18
19      // Release Memory
20      free(source);
21      source = 0;
22
23      free(destination);
24      destination = 0;
25
26      return 0;
27  }
```

All elements of the *source* array are assigned the value 0 using a *for loop* and the array is initialized. Now the second array *destination* is also to be initialized by copying the memory area of the first array into the memory area of the second array.

This is done by the line *memcpy(destination, source, size * sizeof(int));*. The *memcpy* function takes three parameters. The first two parameters specify the destination address and the source address between which the copy operation is to take place. The copy operation occurs from source to destination. The third parameter specifies the number of bytes to copy. In this example, the number of bytes corresponds to the size of the array, but this is not a requirement for the function.

Finally, as in the previous example programs, the reserved memory is released again and the stored addresses are overwritten with the value 0.

11.7 Functions

Functions deal with pointers in several ways. Of course, pointers can be passed as parameters to functions. The behaviour is similar to the *call by reference* described in Sect. 9.3 and is called *call by pointer*.

However, it is also possible to create pointers to functions in order to pass them to another function or class, for example. This makes it possible to dynamically define functions that should be called when a certain event occurs.

These two application purposes will be described in more detail in the following two chapters.

11.7.1 Call by Pointer

When pointers are passed to functions, they initially have a similar effect to the references introduced in Sect. 9.3. Since a pointer passes the address of a memory location, the changes made to the contents naturally cross the boundaries of the function. It is therefore just as possible to pass pointers to have a sub-function change the contents of a variable.

Listing 9.5 was introduced in Sect. 9.3 and can be solved just as well using pointers. The result is shown slightly modified in Listing 11.12.

Listing 11.12 Swapping Two Values Using Pointers

```
1    // A function declaration for a call by pointer
2    void swap(int*, int *);
3
4    // Main function
5    int main ()
6    {
7      // Variable definition and initialization
8      int value1 = 5;
9      int value2 = 10;
10
11     // Transfer of addresses
12     swap (&value1 , &value2);
13
14     return 0;
15   }
16
17   // Definition of the function swap
18   void swap(int* a, int* b)
19   {
20     // Swapping the contents
21     int h = *a;
22     *a = *b;
23     *b = h;
24   }
```

In the function declaration *void swap(int*, int*);* the& symbols are replaced by asterisks to indicate the passed parameters as pointers. Since the two variables *value1* and *value2* were created on the *stack*, the function call *swap(&value1,&value2);* must now be preceded by the ampersand to get the addresses of the variables.

Within the *swap* function, the pointers must be dereferenced by preceding asterisks, as in **a = *b;*, to swap the contents of the pointers. Overall, the use of pointers makes the function more unwieldy than a *call by reference*.

The background is that this is not a situation where a *Call by Pointer* would be used. Basically, the two concepts *Call by Reference* and *Call by Pointer* serve the same purpose. It is also possible to work with both concepts in any situation, but using the wrong concept can make a program unwieldy.

The program should therefore be analyzed in detail. If no pointers are used in the whole program, a *call by reference* is the right choice. If the program only uses pointer variables in relation to the function, a *call by pointer* is the better choice in most cases. If no clear choice can be made, functions with both concepts can also be offered.

11.7.2 Function Pointer

The *C++* language makes it possible to create pointers to functions. This allows functions to be stored and used like normal variables. A typical use case for a function pointer is a so-called *callback function.*

Here, a function of a program or a class executes a certain operation. After the operation is finished, several classes of the software are to be informed about the end of the operation. Of course, it would be possible to write the order of function calls for the respective classes hard into the program. However, it would be much more elegant if an array of function pointers existed. Each class that wants to be informed can then add the address of a function to this array, or remove it again. The number of calls could then be limited to the classes that are really interested in the result.

However, the application just presented is too complicated for the first practical example of a function pointer. Therefore, Listing 11.13 first creates and uses a function pointer for a simple function.

Listing 11.13 Declaring and Using Function Pointers

```
1    #include <iostream >
2
3    using namespace std;
4
5    // Example function
6    int sum(int a, int b)
7    {
8      return a + b;
9    }
10
11   // Example function 2
12   int mul(int a, int b)
13   {
14       return a * b;
15   }
```

```
16
17   int main ()
18   {
19       // Declaration of a function pointer
20       int(* fpointer )(int , int);
21
22       // Value assignment of a function by
23       fpointer = sum;
24       // or
25       fpointer = &sum;
26
27       // The pointer works for all functions
28       // with the same characteristics
29       fpointer = mul;
30
31       // Application of the function pointer
32       cout << fpointer (3, 7) << endl;
33
34       return 0;
35   }
```

Function pointers are always created for a specific type of function. This means that the important features of a function, such as the passing parameters and the return type, must also be specified for the function pointer. By *int sum(int a, int b),* a simple function is defined that returns a value of type *int* and takes two parameters of type *int.* This function calculates the sum of the two passed values.

If another function with the same characteristics is defined, as for example by *int mul(int a, int b),* the function pointer can also be used for this function.

Within the main function, a function pointer is now defined by *int(*fpointer)(int, int);,* which sets the characteristics for the functions it can point to. First, the return type is specified by *int.* The name of the function pointer must be specified with an asterisk inside parentheses. This is followed by the variable types of the function parameters in round brackets. In this example, the pointer can point to a function with return type *int,* which expects two parameters of type *int.*

The value assignment of a function to the pointer is done by *fpointer = sum;.* The name of the function is assigned to the pointer like a normal value. Alternatively, it is also permissible to prefix the name of the function with an ampersand, as in *fpointer = ∑.* However, this is by no means necessary.

The second function can also be used to assign a value by *fpointer = mul;* because it has the characteristics specified for the pointer.

Now the name of the pointer can be used like a function. *cout < <fpointer(3,7)< <endl;* This is shown in the line as an example. The call of the function pointed to by *fpointer*

is done like an ordinary function call and also the return value can be processed normally. In this example, the output of the program is 21.

However, functions often do not create single pointers, instead function pointers to the same function type are often needed in multiple places. If functions with several parameters are involved, errors can easily occur. It is therefore useful to create a new variable type for functions of a certain type. In *C++*, the *typedef* statement exists for defining new variable types. Listing 11.14 shows the use of the *typedef* statement for function pointers.

Listing 11.14 Declaring and Using Function Pointers Using *typedef*

```
1    #include <iostream >
2
3    using namespace std;
4
5    // Example function
6    int sum(int a, int b)
7    {
8       return a + b;
9    }
10
11   typedef int (* fpointer ) (int , int);
12
13   int main ()
14   {
15      // Declaration of a function pointer
16      fpointer fp;
17
18      // Value assignment of a function by
19      fp = sum;
20
21      // Application of the function pointer
22      cout << fp (3, 7) << endl;
23
24      return 0;
25   }
```

In this example, *typedef int(*fpointer)(int, int);* creates a new variable type. The definition of the variable type is the same as the definition of the already familiar function pointer from Listing 11.13. The characteristics of the functions that the pointer can refer to are also identical. However, in this example, the name *fpointer* specifies the name for a variable type rather than a variable.

In the main function, this variable type can now be used to create a function pointer with the name fp using *fpointer fp;*. The value assignment and application of this function pointer is now analogous to Listing 11.13.

Since function pointers can now be used like any other variable type through the *typedef* statement, it is also easy to create arrays of function pointers.

11.8 Classes

The topic classes was already treated in detail in Chap. 10. Nevertheless, it makes sense to open the chapter again after the pointers have been introduced.

So far it has already been shown that the use of pointers is always worthwhile when large memory areas are to be created dynamically. This has led, for example, to the fact that the size of arrays could be freely specified during runtime and even changed subsequently. The possibilities of the arrays were thus extended by the use of pointers.

A similar thing happens with classes. However, a first obvious change is the notation used to access the member functions and variables of the class when it has been created on the *heap*. On p. 133, as part of the description of the *Vector2D* class in Listing, a main function was presented in which an object of the class was created and used. This program is now to be modified so that the object of the class is created on the *heap*. Listing 11.15 shows the new version of the program.

Listing 11.15 The Main Program (Project.cpp)

```
1    #include <iostream >
2    #include "Vector2D.h"
3
4    using namespace std;
5
6    // Main function
7    int main ()
8    {
9      // Variable definition and initialization
10     Vector2D* v1 = new Vector2D ();
11
12     // Auxiliary variables
13     double x = 0.0;
14     double y = 0.0;
15
16     // Output
17     // Value input
18     cin >> x;
19     cin >> y;
20
21     v1 ->setCartesian(x, y);
22
```

```
23       cout << "Vector length: " << v1 ->getLength () << endl;
24
25       // Release Memory
26       delete v1;
27       v1 = 0;
28
29       return 0;
30    }
```

To create a pointer to an object of class *Vector2D*, the original line 10 Vector2D *v1; is* changed to *Vector2D* v1 = new Vector2D();*. The variable *v1* now has the variable type *Vector2D** and an object of the class is created on the *heap* using the *new* statement.

This changes the notation for accessing the class members. Instead of the dot (.), an arrow (–>) is now used. The line 21 *v1.setCartesian(x, y);*, thus becomes *v1–>setCartesian(x, y);*. The same applies to line 23 *cout << "Vector length:" << v1.get-Length() << endl;*, which is changed to *cout << "Vector length:" << v1 –> getLength() << endl;*.

Since the object was created on the *heap*, the memory must be manually released again by *delete v1;*. And of course the address should also be deleted in this example by *v1 = 0;* to prevent incorrect accesses.

Nothing changes within the class declaration. The description of the class itself is independent of where the object is created.

11.8.1 Polymorphism

The full use of polymorphism is actually only possible when pointers are used. If the type of a pointer is a class *A*, then this pointer can also point to objects of classes that inherit from the class *A*. Of course, this pointer can then only be used to call functions that have already been declared in class *A*, but in many cases this is exactly what is needed.

In Listing 11.16, an abstract class is defined to provide an interface for objects to be drawn into the console. To keep the class as small as possible, no *.cpp file* was created. Instead, the definitions of the constructor and destructor were written directly to the *header file*.

Listing 11.16 The Abstract Class *Object* (Object.h)

```
1    // Include Guard
2    class Object
3    {
4    public:
5       Object () {};
6       virtual ~Object () {};
```

```
7
8      virtual void paint () = 0;
9    };
```

It is indeed possible for each function within a class to write the complete definition in the *header* file. However, the clarity of the *header* file suffers in most cases as a result, so it is not advisable to get into the habit of this style. However, for a small abstract class whose constructor and destructor remain empty, this is possible without loss of clarity .

So the line *Object()* { };is the definition of the constructor, where the empty curly braces make it clear that nothing happens in the constructor. Likewise, the line *virtual ~ Object()* {}; defines the destructor. The keyword *virtual* clarifies that *C++* must look in the *vtable* at runtime to see which destructor must be called.

With the last function, which is only declared, the class becomes abstract, because in line 8 *virtual void paint() = 0;* the function *paint* is assigned the value 0. So there is no implementation for this function within the class, consequently the function must be virtual, so that a definition can be added in inheriting classes.

Now two classes are to inherit from the class *Object*. On the one hand the class *Cube*, which is shown in the programs 11.17 and 11.18, and the class *Circle*, whose implementation is in the programs 11.19 and 11.20. Both classes differ only in details, so that a detailed description is carried out only for the class *Circle*.

Listing 11.17 The Inheriting Class *Cube* (Cube.h)

```
1    // Include Guard
2    #include "Object.h"
3
4    class Cube : public Object
5    {
6    public:
7      Cube ();
8      ~Cube ();
9
10     void paint ();
11   };
```

Listing 11.18 The Inheriting Class Cube (Cube.cpp)

```
1    #include "Cube.h"
2    #include <iostream >
3
4    using namespace std;
5
```

```
6    Cube :: Cube ()
7      :Object ()
8    {
9    }
10
11   Cube ::~ Cube ()
12   {
13   }
14
15   void Cube :: paint ()
16   {
17     cout << "****" << endl
18           << "****" << endl
19           << "****" << endl
20           << "****" << endl;
21   }
```

In the *header* file of the *Circle* class, shown in Listing 11.19, the line class *Circle: public Object* first specifies that the *Circle class* should inherit from the *Object class*. The class is to have a constructor and a destructor, as well as an implementation of the *paint* function, which was not defined in the abstract class *Object*.

Listing 11.19 The Inheriting Class *Circle* (Circle.h)

```
1    // Include Guard
2    #include "Object.h"
3
4    class Circle : public Object
5    {
6    public:
7      Circle ();
8      ~Circle ();
9
10     void paint ();
11   };
```

Listing 11.20 The Inheriting *Circle* Class (Circle.cpp)

```
1    #include "Circle.h"
2    #include <iostream >
3
4    using namespace std;
5
```

```
 6   Circle :: Circle ()
 7     : Object ()
 8   {
 9   }
10
11
12   Circle ::~ Circle ()
13   {
14   }
15
16   void Circle :: paint ()
17   {
18     cout << " ** " << endl
19          << "****" << endl
20          << "****" << endl
21          << " ** " << endl;
22   }
```

Within the *.cpp* file, it is specified for the constructor that the constructor of the base class is to be called as the only action. If an object of the *Circle* class is called, the constructor of the base class is executed first, before the constructor of the *Circle* class is executed. This also makes sense, since the base class may have to make configurations that are required in the inheriting class.

The destructor of the *Circle* class should not execute any statement. Nevertheless, after the destructor of the *Circle* class, the destructor of the base class is executed. This is because the destructor in the base class has been declared as virtual. If this were not the case, only one of the two destructors would be called. Which one is called depends on the variable type that is used to delete the corresponding object. More on this in a moment when the main program is explained.

In the *paint* function, only a *cout* statement prints a simple circle to the console.

All instructions and concepts used in programs 11.16 to 11.20 have already been explained in the previous chapters of this book. If you find anything unfamiliar while reading, it is best to refer back to the relevant chapter.

Listing 11.21 now uses all of the previously defined classes to fully explain the concept of polymorphism.

Listing 11.21 The Main Program (Project.cpp)

```
1   #include "Object.h"
2   #include "Cube.h"
3   #include "Circle.h"
4
5   int main ()
```

```
 6   {
 7       // Variable initialization
 8       const unsigned int N = 10;
 9       Object* objects[N];
10
11       // Initialization of the individual objects
12       // of the array depending on the array position
13       for (int i = 0; i < N; i++)
14       {
15         if (i % 2 == 0)
16         {
17         objects[i] = new Cube ();
18         }
19         else
20         {
21         objects[i] = new Circle ();
22         }
23       }
24
25       // Call all paint functions
26       for (int i = 0; i < N; i++)
27       {
28         objects[i]->paint ();
29       }
30
31       // Release Memory for all
32       // array elements
33       for (int i = 0; i < N; i++)
34       {
35         delete objects[i];
36         objects[i] = 0;
37       }
38
39       return 0;
40   }
```

First, the *header* files of the classes are included so that they can be used in the main program. Thereby the file *Object.h* should not have been included, because it was already included by the *header* files of the two inheriting classes.

Two variables are initialized in the main program: First, the constant *N*, which is to determine the size of an array, and the array *objects*. The array *objects* represents an array variant that has not yet been discussed. The array is created on the *stack,* so its size must be constant. However, its elements consist of pointers that will later point to objects that are created on the *heap.* The array has the variable type *Object*,* so it is a pointer to an object of an abstract class that cannot itself be created as an object. The explanation is the same as for the *void pointers*. Since no concrete object is created here, but only a pointer that has a fixed size, it is also possible to create pointers to actually impossible targets, such as something undefined *(void)*, or an abstract class.

Within the *for loop* starting at line 13, the elements of the array are initialized by alternately creating objects of type *Cube* and *Circle* on the *heap* and assigning their addresses to the array elements. This works because both classes have inherited from the class *Object* and thus the classes have a common interface. The abstract function *paint* is implemented in both classes, so that objects of the classes can be created.

In the following loop starting at line 26, the *paint function* is called by each element of the array. This will alternately print cubes and circles to the console. In general, any function defined in the abstract class could now be called, since it must be defined either there or in one of the two inheriting classes. So, with the help of abstract classes and polymorphism, it is possible to define interfaces that all inheriting classes must adhere to. Pointers to the abstract class, or interface, can be used to pass objects to functions, or otherwise process them. However, the functionality of the passed objects depends on the individual implementation within the inheriting class.

This opens up many new possibilities, as action sequences can be further abstracted in this way. However, a little experience and a larger project are necessary to recognize and appreciate the possibilities that arise from this.

The program closes by deleting the memory of all reserved objects from line 33 and overwriting the stored address with the value 0.

11.8.2   Operators

Some operators and their implementations have already been presented in Sect. 10.3. However, some implementations require the use of pointers. For this reason, the example implementations of the operators are continued in this chapter after the necessary terms have been introduced.

Arithmetic Operators

Many of the arithmetic operators can be implemented in a similar way to the examples presented in Sect. 10.3. However, for some of the operators, very individual solutions have been chosen. For example, the increment operator ++ and the decrement operator −−, which increase or decrease the variable value by 1, have two possible notations in C++, which are shown in Listing 11.22.

Listing 11.22 The Main Program (Project.cpp)

```
1   #include <iostream >
2
```

```
3    using namespace std;
4
5    // Main function
6    int main ()
7    {
8      int a = 0;
9      int b = 0;
10
11     cout << a++ << endl;
12     cout << ++b << endl;
13     cout << a << endl;
14     cout << b << endl;
15   }
```

The output of this program is:

```
0
1
1
1
```

The background is that the increment $a + +$ is done after the output by *cout*, while the increment $+ + b$ is done before the output. For the last two outputs, both variables have the value 1.

To distinguish between these two notations in *C++*, a way was chosen that does not seem intuitive at first glance. Listing 11.23 shows the declaration of the two possible increment operators.

Listing 11.23 Declarations of the Decrement Operators (Vector2D.h)

```
1    //...
2
3    Vector2D& operator ++(); // represents ++a;
4    Vector2D operator ++( int); // represents a++;
5
6    // ...
```

The first of the two operators returns a reference to the object of the *Vector2D* class, while the second operator creates a copy of the object and takes an *int* parameter that is used to distinguish the two operators. However, the parameter has no further meaning for the operator itself. The implementation of the two operators is shown in Listing 11.24.

Listing 11.24 Implementing the Decrement Operators (Vector2D.cpp)

```
 1    // ...
 2
 3    // represents ++a;
 4    Vector2D& Vector2D :: operator ++()
 5    {
 6       double v = getLength ();
 7       *this = *this * ((v + 1) / v);
 8       return *this;
 9    }
10
11    // represents a++;
12    Vector2D Vector2D :: operator ++( int)
13    {
14       Vector2D result = *this;
15
16       double v = getLength ();
17       *this = *this * ((v + 1) / v);
18
19       return result;
20    }
21
22    // ...
```

With both operators the length of the vector is to be increased by the value 1. For both operators the length of the vector is calculated by *double v = getLength( ); and* stored in the variable *v.*

To extend the vector by the value 1, formula 11.4 must be applied. For this purpose it is useful to use the already defined operator for scalar multiplication. However, this requires an object of type *Vector2D,* which can be multiplied by the scalar. However, this object was previously only known outside the class.

$$\left| \vec{v} \times \frac{|\vec{v}|+1}{|\vec{v}|} \right| = |\vec{v}|+1 \tag{11.4}$$

Since this problem occurs frequently, the *C*++ language offers each object of a class the possibility to use a pointer to itself. This pointer has the name *this* and always means the object whose function was just called.

The line **this = *this * ((v + 1)/v);* implements formula 11.4. The pointer *this* is dereferenced by the preceding asterisk and thus refers to the data that is hidden behind the pointer. In other words, to the object of the class *Vector2D.* The scalar multiplication operator is applied to this object, multiplying each element of the vector by the value

v $1/v$ $|\vec{v}|$ $1/|\vec{v}|$. The result is again stored in the dereferenced pointer, which is the object itself.

The vector on which the operator ++ was executed was thus extended by the value 1. Now, however, the different behavior expected from both operators must still be mapped.

With the first operator, that is, with $++a$, the modified vector can be returned directly as a result. The line *return *this;* may be a bit confusing, but it makes absolute sense. The pointer *this* points to the object itself and is dereferenced by the preceding asterisk. This returns an object of type *Vector2D*, which has an address, and thus corresponds to the expected return type *Vector2D&*.

The second operator $a++$ must also extend the length of the vector by the value 1, thus uses the same calculation steps for the calculation of the longer vector as the first vector. However, the return value of the operator should be the original vector, whose length has not yet been changed. For this reason, a new object is created in the first line by *Vector2D result = *this;* as a copy of this original vector.

In the following lines the length of the vector is changed and finally the previously stored copy is returned by *return result;*.

Assignment Operators

There are basically two variants of the assignment operators. The first variant combines an arithmetic operation with a value assignment and the second variant directly assigns a specific value to the object.

As an example for the value assignment with combined arithmetic operation serves here the operator +=. The implementations of other operators of this kind depend of course on the respective arithmetic operation, but the basic idea is always the same. To emphasize the differences to the arithmetic operator +, an implementation of this operator is also presented.

The value assignment operator = is different from the assignment operators that perform an arithmetic operation in addition to the value assignment, and has more similarities with the copy constructor. A sample implementation for this operator will also be presented here. Listing 11.25 shows the declarations of the operators from the *header file* of class *Vector2D*.

Listing 11.25 Declarations of the Arithmetic Operator + and the Assignment Operators + = and = (Vector2D.h)

```
1    //...
2
3        // Arithmetic operators
4        Vector2D operator +( Vector2D r) const;
5
6        // Assignment operators
```

```
 7        Vector2D& operator +=( Vector2D r);
 8        void operator =( Vector2D r);
 9
10    // ...
```

Some differences between the two operators + and += can be seen at first glance. The operator + performs a normal vector addition. This means that two vectors $\vec{l}$ and $\vec{r}$ are linked using addition. The result is a new vector $\vec{e}$ of type *Vector2D*. Here, the left operand is the object that performs the operator and the right operand is the function parameter called *r* in the example program. Since this operation does not change the object itself, the operator can be performed on constant objects. The declaration *Vector2D operator + (Vector2D r) const;* can therefore be supplemented by the keyword *const*.

The assignment operator += also performs a vector addition of two vectors $\vec{r}$ and $\vec{l}$. However, the result is not a new vector, but the vector $\vec{l}$ modified by the sum. Since no new object is created, the return type of this operator is *Vector2D&*. In addition, the operator cannot be executed on constant objects because it modifies the object in question.

The assignment operator = creates a direct copy of the passed parameter *r*. Since value assignment does not usually occur within other arithmetic operations, the return type is *void*. The operator cannot be executed on constant objects either, since its only purpose is to change the object on which it is executed.

The implementation of the operators in the *.cpp* file is shown in Listing 11.26.

Listing 11.26 Implementing the Arithmetic Operator + and the Assignment Operators + = and = (Vector2D.cpp)

```
 1    // ...
 2
 3    // Addition with object copy
 4    Vector2D Vector2D :: operator +( Vector2D r) const
 5    {
 6      Vector2D result;
 7
 8      result.m_x = this ->m_x + r.m_x;
 9      result.m_y = this ->m_y + r.m_y;
10
11      return result;
12    }
13
14    // Addition with value assignment
15    Vector2D& Vector2D :: operator +=( Vector2D r)
16    {
17      this ->m_x += r.m_x;
18      this ->m_y += r.m_y;
```

```
19
20      return *this;
21    }
22
23    // Value assignment
24    void Vector2D :: operator =( Vector2D r)
25    {
26      this ->m_x = r.m_x;
27      this ->m_y = r.m_y;
28    }
29
30    // ...
```

First, the normal addition of two vectors is implemented with the operator +. Since the object itself must not be changed, a new object of type *Vector2D* is created by *Vector2D result;* in which the result can be stored.

The result is calculated by the formula 11.5 and is implemented in *C++* by the lines 8 *result.m_x = this −> m_x + r.m_x;* and 9 *result.m_y = this −> m_y + r.m_y;*. For both lines, it would be possible to omit the *this − >* before *m_x* and *m_y,* but this makes it clear that they are the variables of the object itself. The operation + between *this −> m_x* and *r.m_x;* can be applied, since these are variables of type *double*.

$$\vec{c} = \vec{l} + \vec{r} = \begin{pmatrix} l_x \\ l_y \end{pmatrix} + \begin{pmatrix} r_x \\ r_y \end{pmatrix} = \begin{pmatrix} l_x + r_x \\ l_y + r_y \end{pmatrix} \tag{11.5}$$

Finally, *return result;* returns the calculated vector.

If the operation + is combined with the value assignment, as is the case with the operator +=, no new object needs to be created. This makes the operator faster, but it cannot be used on constant objects.

Formula 11.5 is implemented in this operator by lines 17 *this −> m_x += r.m_x;* and 18 *this−> m_y += r.m_y;*, where the result is stored by += directly in the object itself.

As a result of the calculation, a reference to the now changed object is returned by *return *this;*.

The value-only operator only needs to copy the values of the variables of the object to be copied into the variables of the current object. This is done by lines 26 *this −> m_x = r.m_x;* and 27 *this −> m_y = r.m_y;*.

Comparison Operators
Comparison operators relate the current object *A* to another object. For this, of course, a meaningful comparison between the instances of a class must be possible. If the class represents circles, the radius could be used for a size comparison, and for names an alphabetical order would be conceivable. However, there may also be situations where there is no meaningful sorting. For example, if the class represents something abstract, such as

messages that can be sent back and forth between different computers, then a size comparison is probably not possible.

In the example in this book, vectors are compared so that a magnitude comparison can be implemented, and the equality of two vectors is also uniquely defined. Listing 11.27 presents the declarations for two operators, == and >. Since in most cases only two meaningful answers are possible when comparing two objects, the return type of both operators is *bool*.

Listing 11.27 Declarations of the Comparison Operators == and >

```
1   //...
2
3      // Comparison operators
4      bool operator ==( Vector2D r);
5      bool operator >( Vector2D r);
6
7   // ...
```

The implementation of the two operators is shown in Listing 11.28.

Listing 11.28 Implementing the == and > Comparison Operators

```
1    // ...
2
3    // Equality
4    bool Vector2D :: operator ==( Vector2D r)
5    {
6      return (this ->m_x == r.m_x) && (this ->m_y == r.m_y);
7    }
8
9    // Inequality (Larger)
10   bool Vector2D ::operator >( Vector2D r)
11   {
12     return this ->getLength () > r.getLength ();
13   }
14
15   // ...
```

Here, a logical expression is defined in the first operator ==, which checks whether the coordinates of both vectors are identical. In this case the result of the expression is *true*, otherwise *false*.

The second operator compares the magnitudes of both vectors and returns *true* if the magnitude of the left vector is greater than that of the right vector.

Other relational operators can be implemented analogously to this scheme.

11.9 Advanced: Unions

The C++ language contains another variant of the class, which, however, has some special features. The so-called *union* is also a complex structure, which connects several variables to a common data type. However, the *union* uses only as much memory as is needed for the largest member variable, because all variables are located in exactly the same memory area.

Because of this peculiarity, the *Union* has some limitations compared to the normal classes. For example, it is not possible to inherit from *Unions,* or to inherit them. Also the definition of virtual functions is not allowed.

The variables of a *union* must also not be references.

Otherwise, all functions known from normal classes can be defined for *unions,* including constructors, destructors and operators. In most cases, however, a union is only used as a data container.

By unifying the variables contained in the union, a particular record can be accessed in quite different ways. For example, it is possible to define a variable of type *int,* together with an array of four *unsigned chars.* This allows a value to be assigned to the *int variable,* which can be worked with in the normal way. At the same time, however, the array can be used to address each *byte* of the integer value individually. Listing 11.29 implements this example.

Listing 11.29 Implementing a Union Data Structure

```
1   #include <iostream >
2   #include <iomanip >
3
4   using namespace std;
5
6   // Definition of the Union
7   union example
8   {
9     unsigned int value;
10    unsigned char part [4];
11    unsigned char first;
12  };
13
14  int main ()
15  {
16    // Variable declaration
17    example test;
18
19    // Assignment of the value
20      // Hexadecimal fe dc ba 98
```

```
21      test.value = 4275878552;
22
23      // Output of the value as int
24      cout << hex << test.value << endl;
25
26      // Output of the value byte by byte
27      for (int i = 0; i < 4; i++)
28      {
29        cout << (int)test.part[i] << " ";
30      }
31
32      cout << endl;
33
34      // Output of the first byte
35      cout << (int)test.first << endl;
36
37      return 0;
38   }
```

The *union* is defined similarly to a class, only the keyword *class* is replaced by the keyword *union*. Then follows the name of the new data type and within curly brackets the *members* are declared. In this example, three different constructs are to be used.

The largest variable always determines the memory consumption of the *union*. In this case, the variable of type *unsigned int* and the field of type unsigned *char* occupy exactly 4 bytes, while the variable of type *unsigned char* occupies only one byte. The *union* therefore has a size of 4 bytes. The three variables share the same memory within the *union*.

To illustrate what this means, a variable of the type of the *union* was created in the main program by *example test;*. The line *test.value = 4,275,878,552;* initializes the variable *test. value* of the *union* with a value. This value was not chosen randomly, but corresponds exactly to the hexadecimal number *fe dc ba* 98, which therefore covers the full 4 bytes and has a different digit at each position.

First, *cout << hex << test.value << endl;* prints the value of the variable *test.value* to the screen in hexadecimal. This is provided by the keyword *hex,* which is defined in the library *iostream*, and permanently switches the output stream to hexadecimal output.

Within the *for-loop,* the content of the array *test.part* and finally, the content of the variable *test.first* is printed.

The output of the program reads:

```
fedcba98
98 ba dc fe
98
```

First the output of the complete integer value is done, which is *fedcba98* as expected. Then the output is done byte by byte and, as with the pointers, this output is done in reverse order 98 *ba dc fe*. Finally, the first byte of the integer value is printed, the 98.

Exercises

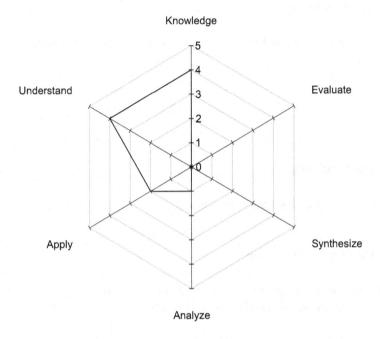

Network diagram for the self-assessment of this chapter

11.1   Memory Areas

Name the four areas into which the memory of a program can be divided!

11.2   Dereferencing

Describe the meaning of the term "dereferencing"!

11.3   Multidimensional Arrays

Explain briefly and in your own words the three ways to create a multidimensional array on the *heap*!

11.4   Function Pointer

Name the instruction that can be used to create function pointers and enumerate the information needed to do so!

11.5   *Stack* and *Heap*

Explain the difference between the stack memory *stack* and the *heap*!

11.6   Memory Consumption

Calculate the amount of memory required by an image that consists of 1024 × 768 pixels and whose pixels encode the color value with 16 bits.

11.7   Pointer Arithmetic

Explain the peculiarity of pointer arithmetic!

11.8   Memory Reservation

Summarize why the *malloc* and *free*, and *new* and *delete* statements should never be used together!

11.9   Random Numbers

Write a program that creates an array *values* with N elements, on the *heap* and initializes it in a *for loop* with random integers in the interval [1; 6]. The number N should be an integer in the interval [1; 1000] and should be chosen by the user.

Calculate the average and standard deviation as in Exercise 8.10 and print the values to the console.

Release the memory at the end.

11.10   Random Numbers the Second

Write a program that generates a two-dimensional array *values* with the dimensions Y = 1000 and X = N, on the *heap*. The number N should be an integer and freely chosen by the user in the interval [1; 10]. Use the third variant, that is, the virtual dimensions, to create the array. Initialize the array using *for loops* with random integers in the interval [1; 6].

Calculate the sum of all values in a row and determine the average and standard deviation from the result, as in Exercise 8.10. Print the values to the console.

Release the memory at the end.

11.11   Program Analysis

Analyze the following program. For once, this program has no deeper meaning, it's just about fiddling a bit with the pointers and their possibilities.

Try to figure out what the output of the program is and explain each line!

Do not type the program, but try to understand what is happening without assistance!

```
1    # include <iostream >
2
3    using namespace std;
4
5    int main ()
6      {
7      int x = 0;
8      int y = 0;
9      int k = 0;
10     double *z = new double (3.0);
11
```

```
12      y = (int)z;
13
14      k = (int )*z;
15
16      x = (int )&y;
17
18      *(( double *)y)  *= 2;
19
20      *z /= k;
21
22      cout << ( int )*(( double *)(*( int *)x)) << endl ;
23
24      delete z;
25      z = 0;
26
27      return 0;
28    }
```

After the basics of the C++ language have been taught in the second part of the book, the knowledge is now to be applied practically by developing a large coherent program that is to fulfill a concrete task. In doing so, the various steps of software development are to be gone through, from the analysis of use cases, through the creation of activity and class diagrams, to the finished program.

The task of electrocardiography was selected as a typical example from medical technology. Here, electrical currents are measured via 12 channels at different positions of the body. The visualized results enable a doctor to draw conclusions about the state of health of the heart muscle.

The 12 channels are composed of the three limb derivations according to Eindhoven (All directions are described from the patient's point of view):

- Derivation I: From the right to the left arm
- Derivation II: From the right arm to the left leg
- Derivation III: From the left arm to the left leg

In addition, there are the three extremity derivations according to Goldberger:

- Derivation aVR: From left arm and leg to right arm
- Derivation aVL: From left leg and right arm to left arm
- Derivation aVF: From right and left arm to left leg

And the six Wilsonian chest wall leads:

B. Tolg, *Computer science to the Point*,
https://doi.org/10.1007/978-3-658-38443-2_12

Fig. 12.1 Heartbeat

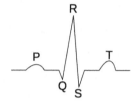

- V1: Located in the fourth intercostal space, counting from the clavicle, on the right side of the sternum.
- V2: Placed in the fourth intercostal space, counting from the clavicle, on the left side of the sternum.
- V3: Is placed exactly between V2 and V4, i.e. on the fifth rib.
- V4: Placed in the fifth intercostal space, counting from the clavicle, so that it is exactly centered under the left clavicle.
- V5: Is placed at the height of V4, so that it lies exactly under the highest point of the left-hand axis fold.
- V6: Is placed at the level of V4 so that it lies exactly under the middle of the left armpit.

The Eindhoven derivative II ideally generates a course for each heartbeat, as shown in Fig. 12.1. The individual phases of the heartbeat are designated in medicine by the letters P, Q, R, S and T and represent different cardiac activities. This constantly recurring course of the ECG data during a heartbeat with its various phases makes it possible to recognize cardiac activity with the aid of pattern recognition algorithms. Ideally, the R wave is particularly easy to recognize and, if the distances between different R waves are measured, provides the possibility to determine the heart rate.

However, various diseases change the course of the data, sometimes dramatically, so that detection of the processes in the general case is a non-trivial problem.

This brief introduction should make it possible to roughly interpret the meaning of the individual leads and the resulting data. A complete introduction to the complex topic of ECG can be found in Gertsch (2008). In this book, the focus will be on the development of software that can read and analyze ECG data.

12.1 Planning the Software Architecture

Before the software is developed, it should first be analyzed what the later software is supposed to do. An unplanned change in the basic structure of the software usually becomes more expensive the later it occurs in the development. For this reason, the use cases planned for the software should be examined first.

Since even a simple program can quickly become very complex, only the three use cases shown in Fig. 12.2 will be considered for this example.

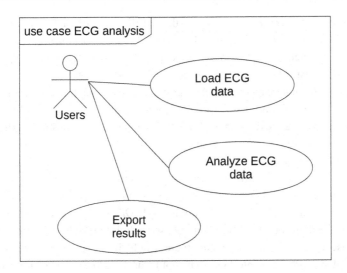

Fig. 12.2 Use case diagram for ECG analysis

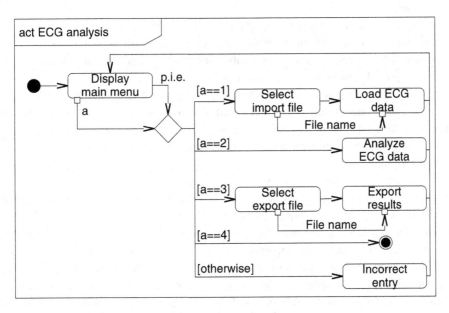

Fig. 12.3 Activity diagram for the ECG analysis user interface

The user should be able to read ECG data with the help of the program, analyse it to a limited extent and export the results.

These three use cases can already be examined and described in more detail. The later program needs a kind of main menu, with the help of which a selection can be made between the three use cases. The details can be documented in an activity diagram. Figure 12.3 shows a possible flow that allows a selection between the use cases.

In principle, there is no specification in the development of such diagrams as to how large or complex they must be designed. Nevertheless, it makes sense, especially at the beginning, to approach a problem in several steps. The activity diagram in Fig. 12.3 therefore deliberately abstracts some steps in order to describe them more precisely in further diagrams. Although this results in many diagrams, these are small and clear in themselves. In addition, at the beginning of software development, all problems must first be thought through. When creating the diagrams, it may then become apparent that certain cases have not yet been considered. Many small diagrams can then be changed more easily than a large canvas.

However, at a later stage of the software development it is quite useful to create a large overview graphic.

The process in Fig. 12.3 starts at the activity node *Display main menu*. In principle, this would have to be described in more detail, but we will not do so here.

In the case of custom development, there would be concrete requirements for the appearance of the program interface that would have to be planned and implemented exactly. Here, however, it is first a matter of finding out how the program must be structured in general.

The main menu is a simple text output that is placed in a class that is responsible for user interaction. There obviously needs to be some user input that generates an integer value a, which is used to choose between the various options. A separate activity diagram would not provide any additional information here.

The object and control flow of the activity node *Representation Main Menu* is routed to a decision node after the activity is completed. Based on the value of a, this redistributes the control flow into one of four different flows.

The simplest case occurs for $a == 4$. In this case, the activity and thus the entire program is terminated.

In all other cases, various activity nodes are traversed, which always lead back to the representation of the main menu at the end. However, the various activity nodes differ greatly in their complexity. Should a value be entered for which no action is defined, an error message must be issued and returned to the main menu.

In the event that data is to be imported or exported, a file must of course first be selected. For this purpose, there are two activity nodes *Select import file* and *Select export file* that perform this task. Both nodes must generate a file name that can be passed on to the following nodes. The detailed description of these two activity nodes can be created without further research and is given in Figs. 12.4 and 12.5.

The remaining three activity nodes *Load ECG data, Analyze ECG data,* and *Export results* are much more complicated, as no information is yet available about the data format or how to analyze the data. This investigation takes place in three separate subchapters.

The two activities *Select import* or *export file* are very similar in their basic structure. First, a file path must be selected for import or export, which is then forwarded to a file selection. The *File Selection* activity node in both diagrams refers to the same activity

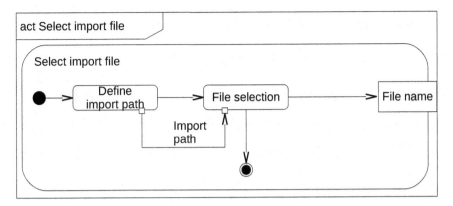

Fig. 12.4 Activity diagram for selecting an import file

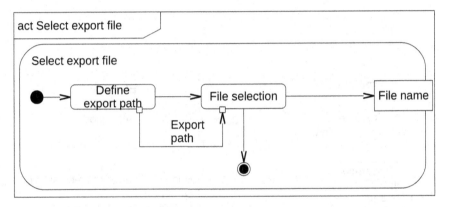

Fig. 12.5 Activity diagram for selecting an export file

described in Fig. 12.6. A separate diagram can also be created for the selection of the path for import and export, in which it can be described how exactly the selection is to be made. For this example, a fixed file path is to be selected in both cases, which is passed to the file selection. Thus, no further diagrams are required for these activity nodes.

The file selection is to take place via the console. The path moves via an object flow in the first activity node *User input*. There the user gets the possibility to enter a file name. This is completed with the path and forwarded to the output pin, after which the activity is ended.

With the previous considerations, a class diagram can already be created. Once again, the *Model-View-Controller* design pattern is used as the basis. Figure 12.7 shows the first draft of the class diagram.

All tasks discussed so far are assigned to a class *Frontend*, which is responsible for user interaction. The complete management of the ECG data is done by the class *ECG-Data*.

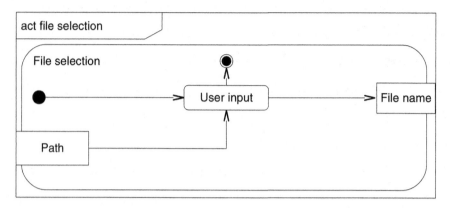

Fig. 12.6 Activity diagram for file selection

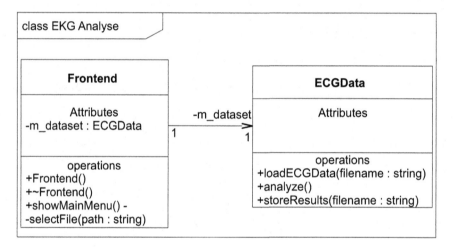

Fig. 12.7 First draft for a class diagram for ECG analysis

The class *Frontend* has exactly one object *m_dataset* of the class *ECG-Data*. In addition, the class has a constructor and a destructor, a function *showMainMenu*, which is responsible for the main menu control and the auxiliary function *selectFile*, in which the file selection is to be implemented.

The class *Frontend* can already be realized as a program, since all further tasks must be realized within the class *ECG-Data*. This class can currently only be defined as a frame. In the following chapters the contents of the class will be examined in more detail.

Listings 12.1 and 12.2 show the translation of the previous diagrams into the language *C++*.

Listing 12.1 The Frontend class (Frontend.h)

```
 1 #include "ECGData.h"
 2 #include <string >
 3
 4 using namespace std;
 5
 6 class Frontend
 7 {
 8 public:
 9 Frontend ();
10 Frontend ();
11
12 void showMainMenu ();
13
14 private:
15 string selectFile(string filter );
16
17 ECGData m_dataset;
18 };
```

Listing 12.2 The Frontend class (Frontend.cpp)

```
 1 #include "Frontend.h"
 2 #include "windows.h"
 3 #include <iostream >
 4 #include <filesystem >
 5
 6 using namespace std;
 7
 8 Frontend :: Frontend ()
 9 {
10 }
11
12
13 Frontend ::~ Frontend ()
14 {
15 }
16
17 void Frontend :: showMainMenu ()
18 {
19 // Variable definition and initialization
20 int a = 0;
21 string path = "";
```

```
22 string filename = "";
23
24 for (;;)
25 {
26 // Main menu display
27 system ("cls ");
28
29 cout << "Welcome to the ECG "
30 << "analysis program !" << endl;
31 cout << "You have the following options "
32 << "at disposal :" << endl;
33 cout << "1: Load ECG data" << endl;
34 cout << "2: Analyze ECG data" << endl;
35 cout << "3: Export results" << endl;
36 cout << "4: End program" << endl << endl;
37 cout << "Please make your selection :"
38 << endl;
39 cin >> a;
40
41 // Decision node
42 switch (a)
43 {
44 case 1: // [a==1]
45 // Select import file
46 path = "C:\ import \\"; // Set import path
47 filename = selectFile(path ); // File selection
48
49 // Loading ECG data
50 m_dataset.loadECGData(filename );
51 break;
52 case 2: // [a==2]
53 // Analyzing ECG data
54 m_dataset.analyze ();
55 break;
56 case 3: // [a==3]
57 // Select export file
58 path = "C:\ export \"; // Set export path
59 filename = selectFile(path ); // File selection
60
61 // Export results
62 m_dataset.storeResults(filename );
63 break;
64 case 4: // [a==4]
65 return;
66 break;
```

```
67 default: // [else]
68 cout << endl << "unknown input !"
69 << endl << endl;
70 cin.get ();
71 break;
72 }
73 } // Return to the main menu
74 }
75
76 string Frontend :: selectFile(string path /* input pin */)
77 {
78 // User input
79 string filename = "";
80
81 cout << "Please enter a filename:"
82 << endl;
83 cin >> filename;
84
85 // Data for output pin
86 return path + filename;
87 }
```

The *header file* is a direct translation from the class diagram 12.7. All operations and attributes have been declared with their respective visibility levels as member variables and member functions in the class.

The definitions of the functions use the information from the various activity diagrams. The comments within the program always refer to the names of the activity nodes that were realized at this point.

12.2   Loading the ECG Data

In order to be able to load ECG data, it must first be researched in which format the data is usually stored and which information is stored in the files. Then a decision must be made as to which file formats are to be supported.

A search on the Internet showed that the website *Physionet* (Goldberger et al. 2000) provides a database with different physiological data sets. More precisely, the database of the Physikalisch-Technische Bundesanstalt (PTB) (Bousseljot et al. 1995) will be used for this book.

All datasets available there are stored in the MIT format *.dat*, the specification of which is also available via *Physionet* (Moody 2018). In addition, there is a software package that can be used to access these data, the so-called *WFDB*.

12.2.1 Description of the MIT Format

The MIT format divides the stored data into several files. The header file, which has the extension *.hea,* contains general information in text format. Among other things, it contains information on how many data records a recorded channel consists of and in which file this data can be found.

Comments can be introduced in any line by the hash character (#) at the beginning of the line. For this example, these lines are to be collected in a *string.*

The first line of the *header file,* the so-called *record line,* contains general information for all data records to which the file refers. Figure 12.8 shows its structure as a syntax diagram.

The complex structure of this line arises from the fact that many elements are optional and can only appear under certain circumstances. The elements mean in detail:

- **Record name:** The name of the data record, can consist of letters, numbers and underscores. Data type: ***string.***
- **Number of segments:** Data records can consist of several segments. In this case, the *record line* would be followed by *segment specification lines.* Otherwise, *signal specification lines* would follow. In this book, segmented files are not supported, so files with this field could not be loaded. Data type: ***int.***
- **Signal count:** The number of recorded channels. Data type: ***int.***
- **Sample rate:** Indicates the number of samples per second per signal. Data type: ***double.***

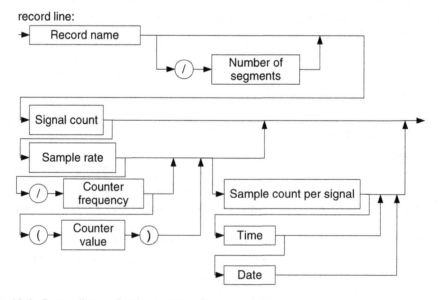

Fig. 12.8 Syntax diagram for the structure of the *record line*

- **Counter frequency:** The counter frequency is used by the *WFDB* software package to transfer strings that are in the so-called standard time format into the time of the data recording. If the value is not specified, it corresponds to the sample frequency. This value is to be ignored by the software described here. Data type: ***double***.
- **Counter value:** The value defines the counter value to be assumed at the first sample. If it is not present, the value 0 is assumed. This value is also to be ignored by the software described here. Data type: ***double***.
- **Sample count per signal:** Specifies how many samples a signal consists of. Is considered undefined if the value is 0 or not present. Data type: ***int***.
- **Time:** Start time of the recording in the format HH:MM:SS with a 24-hour time specification. If the time is not specified, 0:0:0 is assumed. Data type: ***string***.
- **Date:** The date of the recording in the format DD/MM/YYYY. Data type: ***string***.

The *record line* is usually followed by several lines containing information for each recorded signal or segment. Since the segments are to be ignored for the example in this book, they are always followed by so-called *signal specification lines,* the structure of which is shown in the syntax diagram in Fig. 12.9.

The elements of this line mean:

- **File** name: The name of the file in which the signal was saved. Several signals can be combined in one file, so it is important to read this value in each line. Data type: ***string***.
- **Format:** The format specifies how many bits were used to store the signal amplitude. Possible values are: 8, 16, 24, 32, 61, 80, 160, 212, 310, 311. In this book only the format 16 shall be supported. Data type: ***int***.
- **Sample Factor:** Individual signals may have a higher sampling rate than others. In this case, the sample factor indicates the number of samples that were stored instead of a normal sample. The default value is 1. Data type: ***unsigned int***.
- **Offset:** In most cases all signals of a recording are synchronous. The offset offers the possibility to specify the number of additional samples for individual signals whose recording has already started earlier. The value is always positive and indicates how many additional samples were recorded before the first common sample of all signals. The default value is 0. Data type: ***int***.
- **Byteoffset:** The byte offset specifies the number of bytes from the beginning of the file to the first sampled data. The value is normally 0 and is not created by the *WFDB*. Data type: ***int***.
- **AD gain:** To transform analog signals into digital data, an analog-to-digital (AD) converter is used. The AD gain specifies how much the digitized value changes when the recorded analog signal changes by the value of one unit. If the value is 0, or is not specified, the default value is 200. data type: ***double***.
- **Base value:** The base value specifies the sample value that would correspond to the value 0 of the analog data. If it is not specified, it is assumed to correspond to the AD zero value. Data type: ***int***.

Signal specification line:

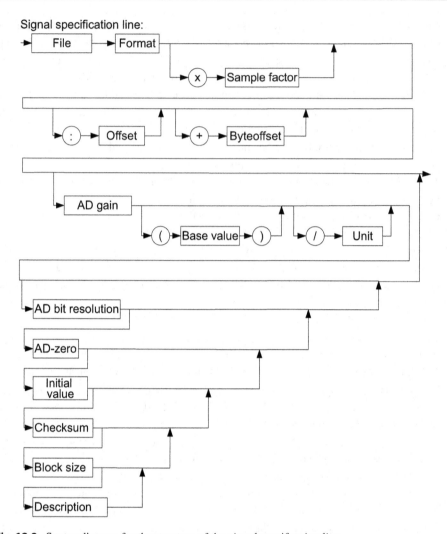

Fig. 12.9 Syntax diagram for the structure of the *signal specification line*

- **Unit:** A text describing the unit of the analog signal. If the unit is not specified, the default value is *mV* Data type: ***string***.
- **AD bit resolution:** Specifies the bit resolution of the digitized data. If the value is missing, the value 12, 10 or 8 is assumed depending on the selected format. For the example in this book, the default value is 12. Data type: ***int.***
- **AD-zero:** The AD-zero value corresponds to the sample value that exactly matches the average value of the possible analog input signal of the AD-converter used. Data type: ***int***.
- **Initial value:** Specifies the value of the first sample of the recording. Certain formats do not store the absolute values, but always the difference to the previous value. In this

case, an initial value is required. The format 16 used in this example does not need this value. The value is assumed to be AD zero if it does not exist. Data type: *int*.

- **Checksum:** A 16 bit checksum of all samples in the recording. This value can be used to check if the values in the file have been corrupted. This value is ignored in this example. Data type: *int*.
- **Block size:** Specifies whether the data must be read in blocks of the specified size. The default value, which is also assumed in this book, is 0. Data type: *int*.
- **Description:** A text that describes the stored signal. Data type: *string*.

In addition to the *header file,* there are two other files. The *.dat file* and the *.xyz file*, both of which contain the stored signals. The *format 16* stores the values in 16-bit format. The two's complement is used, the *least significant* byte is at the beginning and sign bits are multiplied to fill the 16 bits if necessary.

12.2.2 Extension of the Software Architecture

With the information from the file format specification, the software architecture can be further developed. First, the activity diagram for loading ECG data shown in Fig. 12.10 can be developed.

The MIT file format splits into several files, a *header file* and several data files, which are described in more detail in the *header file*. Consequently, the header *file* must first be opened in an activity node in order to obtain all further information.

Of course, this can fail for a number of reasons. For example, there could be an error in the file name, or the file could be on a removable disk that has since been removed. In any case, a decision node must be used to check whether the file was loaded successfully. To do this, the object flow is split. One object token moves to the decision node, another to the next activity node, where it can be processed further. If the loading of the file failed, a message should appear on the screen, otherwise the file can be read out.

The individual lines of the *header file* follow a certain structure, which was presented in Figs. 12.8 and 12.9. This structure must be analysed by the software and broken down into information that is easier for the program to understand. A software component that does this is called a *parser.* The process itself is called *parsing*.

In this case, two different pieces of information are generated during this process. First of all, it can always be evaluated whether the *parsing* was successful. In the event of an error, a message can be issued.

In addition, the *header file* describes which signals were recorded and how they are distributed among the individual files. A set of signals located within a file is called a *group*. This information is needed for further processing so that the correct data can be extracted from the files.

If the *parsing* of the *header file* was successful, the file with the data for each *group* must now be opened and the associated signals read out. This process is basically very

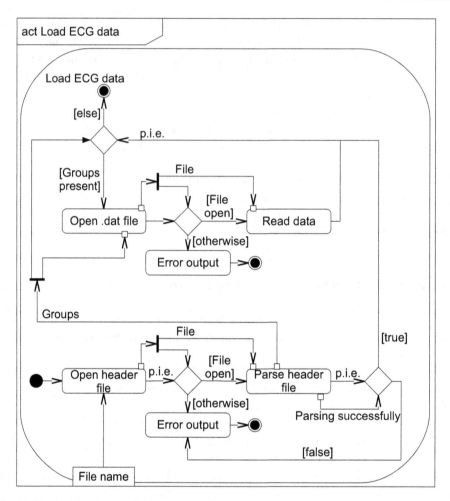

Fig. 12.10 Activity diagram for importing ECG data

similar to reading the *header file*, even though the *Read data* activity node is very different from the *Parse header file* activity node.

However, this process must be repeated as long as there are still groups whose data must be read in. When all data of all groups have been read in, the activity can be successfully completed.

There are some nodes in this diagram that could be described in more detail. However, this will not be done here, as the file format is already very comprehensively documented by the syntax diagrams and the specification (Moody 2018).

In addition to the new activity diagram, the class diagram in Fig. 12.7 can now also be extended. Through the specification, new knowledge about the structure of the data structure could be gained. This new information has been incorporated into the class diagram in Fig. 12.11.

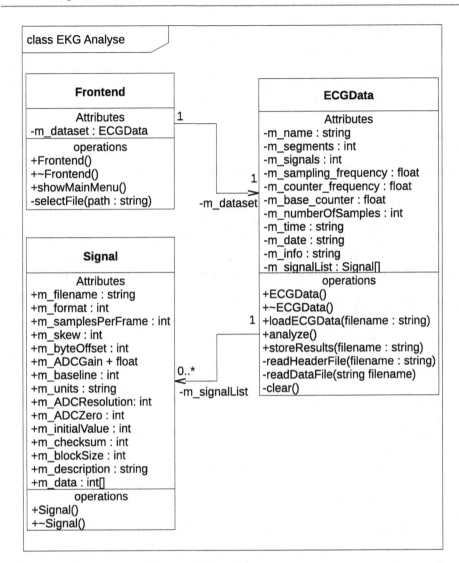

Fig. 12.11 First further development of the class diagram for ECG analysis

A new class *Signal* was created, in which the information of one recorded signal can be stored. The attributes of this class are mainly composed of the information of the *signal specification line* shown in Fig. 12.9. These attributes are supplemented by an array of *int values* in which the recorded data can be stored. All attributes of this class have been assigned the visibility level *public,* since this class only serves as a data container for the class *ECGData* and, apart from the value initialization, has no logic of its own.

The class *ECGData* can now also be filled with more content. The attributes of this class are largely composed of the information of the *record line* shown in Fig. 12.8. The attributes are supplemented by a *string m_info,* in which the comments of the *record line* are to be stored, and by a field *m_signalList* of type *Signal[ ],* in which the list of the loaded signals is stored.

In addition, operations are added that enable the reading of *header* and data files. Furthermore, an operation *clear( )* is added, with which the attributes of the class are to be reset to the initial state.

During the actual implementation it will become apparent that it may be useful to add further auxiliary functions. However, these functions would contribute neither to the understanding nor to the overview in this diagram.

12.2.3 Implementation of the Loading Function

The next step is to implement the new functions and classes. For this purpose, the new class *Signal* is to be created first. Listing 12.3 shows the *header file of* the class.

Listing 12.3 The Signal class (Signal.h)

```
 1 #include <string >
 2 #include <vector >
 3 using namespace std;
 4
 5 class Signal
 6 {
 7 public:
 8 Signal ();
 9 Signal ();
10
11 // Signal Specification Line Data
12 string m_filename;
13 int m_format;
14 int m_samplesPerFrame;
15 int m_skew;
16 int m_byteOffset;
17 double m_ADCGain;
18 int m_baseline;
19 string m_units;
20 int m_ADCResolution;
21 int m_ADCZero;
22 int m_initialValue;
23 int m_checksum;
24 int m_blockSize;
25 string m_description;
26
```

```
27 // Signal data
28 vector <int> m_data;
29 };
```

Most function and variable declarations are made as described in the second part of this book. However, the variable *m_data* has the previously unknown variable type *vector < int >*.

This variable type represents a class implemented in the *C++* Standard Template Library, called *STL*. In this library there are some very useful classes, which simplify the implementation of many problems significantly.

The class *vector* has the properties of an array, but additionally offers the possibility to append new values to the array at any time (*m_data.push_back(5);*) or to query its current size (*m_data.size();*). The name of the variable type is followed by additional information in angle brackets. The array should consist of variables of type *int*. Individual elements of the *vector* are accessed, as with any other array, by specifying the index in square brackets (*m_data[0];* returns the first element, for example).

If the data type to be processed in a class can be freely selected within angle brackets, then it is a so-called template class. This explains the name of the standard template library.

Listing 12.4 shows the implementation of the definition of the *Signal* class, which essentially consists of a constructor and a destructor.

Listing 12.4 The Signal class (Signal.cpp)

```
 1 #include "Signal.h"
 2
 3 Signal :: Signal ()
 4 : m_filename ("")
 5 , m_format (16)
 6 , m_samplesPerFrame (1)
 7 , m_skew (0)
 8 , m_byteOffset (0)
 9 , m_ADCGain (200)
10 , m_baseline (0)
11 , m_units ("mV")
12 , m_ADCResolution (12)
13 , m_ADCZero (0)
14 , m_initialValue (0)
15 , m_checksum (0)
16 , m_blockSize (0)
17 , m_description ("")
18 {
19 }
20
```

```
21
22 Signal :: Signal ()
23 {
24 m_data.clear ();
25 }
```

Within the constructor, all member variables are assigned default values taken directly from the MIT format specification. Within the destructor, the *clear()* function of the *vector class* is called, which deletes all elements of the field. This call is not strictly necessary for values of type *int,* since the function call does not cause the *vector* to free the memory it occupies. This only happens when the object *m_data* ceases to exist, i.e. a short time later.

However, if there are variables with a different data type stored in the vector, such as custom classes, calling the *clear()* function will ensure that their destructors are called and thus given the opportunity to free their memory. For this reason, it makes sense if freeing all array elements is done as a matter of principle as a good practice.

The *ECGData* class is now provided with attributes, operations, and logic for the first time. The implementation of the *header file* is shown in Listing 12.5.

Listing 12.5 The ECGData class (ECGData.h)

```
 1 #include "Signal.h"
 2 #include <string >
 3 #include <vector >
 4 #include <map >
 5
 6 using namespace std;
 7
 8 class ECGData
 9 {
10 public:
11 ECGData ();
12 ECGData ();
13
14 // Main access functions for the
15 // three use cases
16 void loadECGData(string filename );
17 void analyze ();
18 void storeResults(string filename );
19
20 private:
21 void clear ();
22
23 // Reading in the header file
24 bool readHeaderFile(string filename );
```

```
25
26 // Auxiliary functions for disassembling a
27 // Line into individual text sections
28 bool parseLine(string line );
29 vector <string> getChunks(string line );
30
31 // Auxiliary functions for reading in a
32 // record line
33 bool parseRecordLine(vector <string> chunks );
34 bool parseRecordName(string chunk );
35 bool parseSignals(string chunk );
36 bool parseSamplingFrequency(string chunk );
37 bool parseNumberOfSamples(string chunk );
38 bool parseTime(string chunk );
39 bool parseDate(string chunk );
40
41 // Auxiliary functions for reading in a
42 // Signal Specification Line
43 bool parseSignalSpecificationLine
44 (vector <string> chunks );
45 bool parseFileName(string chunk );
46 bool parseFormat(string chunk );
47 bool parseADCGain(string chunk );
48 bool parseADCResolution(string chunk );
49 bool parseADCZero(string chunk );
50 bool parseInitialValue(string chunk );
51 bool parseChecksum(string chunk );
52 bool parseBlockSize(string chunk );
53 bool parseDescription(string chunk );
54
55 // Reading in the data files
56 bool readDataFile(string filename );
57
58 // Auxiliary variables for loading the data
59 vector <Signal> m_signalList;
60 map <string , vector <int >> m_groups;
61 bool m_recordLine;
62
63 // Data of the record Line
64 string m_name;
65 int m_segments;
66 int m_signals;
67 double m_sampling_frequency;
68 double m_counter_frequency;
69 double m_base_counter;
```

```
70 int m_numberOfSamples;
71 string m_time;
72 string m_date;
73 string m_info;
74 };
```

The *ECGData* class essentially declares in the *header file* the functions already presented in Fig. 12.11. However, in order to read the information of the MIT header file, some auxiliary functions have been added. The *parseLine* and *getChunks* functions break a read line into small, contiguous sections that can be more easily parsed. The auxiliary functions, whose names begin with *parse* … then analyze these subsections in detail.

In addition to the already known *vector,* another new data type from the STL is found in the auxiliary variables for reading in the data. The data structure *map* also has many properties of an array, but adds another new feature. Two variable types can be defined within the angle brackets, in this case a *string* and a *vector* ⟨*int*⟩. The first of the two data types specifies the form in which the indices of the array are to be stored, the second data type specifies the type of data stored. So in this case an array of *int values* is assigned to a text.

The concept is to be clarified with an example. First the signal information from the MIT header file is read into the array m_signalList. At the same time, the variable *m_groups* is used to store which signals are assigned to a particular file. Table 12.1 shows how the storage should work.

Table 12.1 Information storage of signal data

m_signalList	
Index	Signal
0	I
1	II
2	III
3	aVR
4	aVL
5	aVF
6	V1
7	V2
8	V3
9	V4
10	V5
11	V6
12	vx
13	vy
14	vz
m_groups	
File	Signal index
data.dat	0, 1, 2, 3, 4, 5, 6, 7, 8, 9, 10, 11
data.xyz	12, 13, 14

The upper table shows the signals, in the order they were defined in the MIT header file at the respective index positions 0–14. The lower table shows that signals 0–11 are assigned to the *data.dat* file, while signals 12–14 are assigned to the *data.xyz* file. These last three signals complete the 12-lead ECG with the three leads according to Frank (Bousseljot et al. 1995; Gertsch 2008).

The remaining attributes correspond to the data stored in the *record line.*

The following Listing 12.6 shows the definition of the class *ECGData.* It was refrained from splitting the source code into small sections to allow a coherent overview of the program. The description of the individual functions follows the program.

Listing 12.6 The ECGData class (ECGData.cpp)

```
 1 #include "stdafx.h"
 2 #include "ECGData.h"
 3 #include <iostream >
 4 #include <fstream >
 5
 6 using namespace std;
 7
 8 // Constructor
 9 ECGData :: ECGData ()
10 {
11 // The clear () function is used for both
12 // the deletion of data used ,
13 // as well as for the initialization of the class
14 clear ();
15 }
16
17 // Destructor
18 ECGData ::~ ECGData ()
19 {
20 clear ();
21 }
22
23 // Initialization of the variables
24 void ECGData :: clear ()
25 {
26 m_recordLine = false;
27 m_name = "";
28 m_segments = 0;
```

```
29 m_signals = 0;
30 m_sampling_frequency = 250;
31 m_counter_frequency = 250;
32 m_base_counter = 0;
33 m_numberOfSamples = 0;
34 m_time = "";
35 m_date = "";
36 m_info = "";
37
38 m_signalList.clear ();
39 m_groups.clear ();
40 }
41
42 // Conversion of the activity diagram Load ECG data
43 void ECGData :: loadECGData(string filename)
44 {
45 // Open header file and
46 // Decision node
47 if (! readHeaderFile(filename ))
48 {
49 // deleting data that may already have been recorded
50 clear ();
51
52 // Activity node error output
53 cout << "Failed to load header !"
54 << endl << endl;
55 cin.get ();
56
57 return;
58 }
59
60 // Open .dat file and
61 // Decision node
62 if (! readDataFile(filename ))
63 {
64 // delete any data already recorded
65 clear ();
66
67 // Activity node error output
68 cout << "Failed to load data !"
69 << endl << endl;
70 cin.get ();
71
72 return;
73 }
```

```
74
75 cout << m_info << endl;
76
77 cin.get ();
78 }
79
80 // not yet implemented
81 void ECGData :: analyze (){}
82
83 // not yet implemented
84 void ECGData :: storeResults(string filename ){
85 }
86
87 bool ECGData :: readHeaderFile(string filename)
88 {
89 // delete already existing data
90 clear ();
91
92 fstream file;
93 string line = "";
94
95 // Open header file
96 file.open(filename , ios::in);
97
98 // Decision node
99 if (file.is_open ())
100 {
101 // Repeat until the end of the file
102 // (end of file - eof)
103 // was not reached
104 while (! file.eof ())
105 {
106 // Reading a line
107 getline(file , line );
108
109 // Analyzing the line
110 // Decision node
111 if (! parseLine(line ))
112 {
113 // Closing the file
114 file.close ();
115 return false;
116 }
117 }
118 // Closing the file
```

```
119 file.close ();
120 }
121 else
122 // Abort , if the file
123 // could not be opened
124 return false;
125
126 return true;
127 }
128
129 bool ECGData :: parseLine(string line)
130 {
131 // If the file starts with a # ,
132 // or empty
133 if (line.find('#') == 0 || line.length () == 0)
134 {
135 // Save as comment
136 m_info += line + "\n";
137 }
138 else
139 {
140 if (m_recordLine)
141 {
142 // should be a Signal Specification Line
143 if (! parseSignalSpecificationLine
144 (getChunks(line )))
145 return false;
146 }
147 else
148 {
149 // If no record line yet
150 // was analyzed
151 // it should be a record line
152 if (! parseRecordLine(getChunks(line )))
153 return false;
154 m_recordLine = true;
155 }
156 }
157 return true;
158 }
159
160 vector <string> ECGData :: getChunks(string line)
161 {
162 int pos = 0;
163 string chunk = "";
```

```
164 vector <string> chunks;
165
166 // as long as the line still contains characters
167 while (line.length () > 0)
168 {
169 // Finding the last space
170 pos = line.find_last_of(' ');
171
172 if (pos != string :: npos)
173 {
174 // if there was a space
175 // the array is filled with the part
176 // after the space
177 chunks.push_back(
178 line.substr(pos + 1, string :: npos ));
179 // and is subsequently shortened
180 line = line.substr(0, pos);
181 }
182 else
183 {
184 // without spaces the remaining
185 // line taken over
186 chunks.push_back(line );
187 line = "";
188 }
189 }
190
191 return chunks;
192 }
193
194 bool ECGData :: parseRecordLine(vector <string > chunks)
195 {
196 int pos = chunks.size () - 1;
197
198 // analyzes the individual sections of the
199 // record line these are now available as parts in
200 // the array chunks
201 if (pos >= 0)
202 {
203 if (! parseRecordName(chunks[pos ]))
204 return false;
205 pos --;
206 }
207 if (pos >= 0)
208 {
```

```
209 if (! parseSignals(chunks[pos ]))
210 return false;
211 pos --;
212 }
213 if (pos >= 0)
214 {
215 if (! parseSamplingFrequency(chunks[pos ]))
216 return false;
217 pos --;
218 }
219 if (pos >= 0)
220 {
221 if (! parseNumberOfSamples(chunks[pos ]))
222 return false;
223 pos --;
224 }
225 if (pos >= 0)
226 {
227 if (! parseTime(chunks[pos ]))
228 return false;
229 pos --;
230 }
231 if (pos >= 0)
232 {
233 if (! parseDate(chunks[pos ]))
234 return false;
235 pos --;
236 }
237
238 return true;
239 }
240
241 bool ECGData :: parseRecordName(string chunk)
242 {
243 char name [2048];
244 // analyzes the format of the record name
245 sscanf_s(chunk.c_str ()
246 , "%s/%i"
247 , &name
248 , sizeof(name)
249 , &m_segments );
250
251 m_name = name;
252
253 return true;
```

```
254 }
255
256 bool ECGData :: parseSignals(string chunk)
257 {
258 // analyzes the format of the number of signals
259 sscanf_s(chunk.c_str(), "%i", &m_signals );
260
261 return true;
262 }
263
264 bool ECGData :: parseSamplingFrequency(string chunk)
265 {
266 // analyzes the format of the sample rate
267 sscanf_s(chunk.c_str ()
268 , "%f/%f(%f)"
269 , &m_sampling_frequency
270 , &m_counter_frequency
271 , &m_base_counter );
272
273 return true;
274 }
275
276 bool ECGData :: parseNumberOfSamples(string chunk)
277 {
278 // analyzes the format of the number of samples per signal
279 sscanf_s(chunk.c_str(), "%i", &m_numberOfSamples );
280
281 return true;
282 }
283
284 bool ECGData :: parseTime(string chunk)
285 {
286 // analyzes the format of time
287 m_time = chunk;
288
289 return true;
290 }
291
292 bool ECGData :: parseDate(string chunk)
293 {
294 // analyzes the format of the date
295 m_date = chunk;
296
297 return true;
298 }
```

```
299
300 bool ECGData :: parseSignalSpecificationLine
301 (vector <string> chunks)
302 {
303 int pos = chunks.size () - 1;
304
305 m_signalList.push_back(Signal ());
306
307 // analyzes the individual sections of the signal
308 // Specification line
309 // these are now as parts in the array chunks
310 if (pos >= 0)
311 {
312 if (! parseFileName(chunks[pos ]))
313 return false;
314 pos --;
315 }
316 if (pos >= 0)
317 {
318 if (! parseFormat (chunks[pos ]))
319 return false;
320 pos --;
321 }
322 if (pos >= 0)
323 {
324 if (! parseADCGain(chunks[pos ]))
325 return false;
326 pos --;
327 }
328 if (pos >= 0)
329 {
330 if (! parseADCResolution(chunks[pos ]))
331 return false;
332 pos --;
333 }
334 if (pos >= 0)
335 {
336 if (! parseADCZero(chunks[pos ]))
337 return false;
338 pos --;
339 }
340 if (pos >= 0)
341 {
342 if (! parseInitialValue (chunks[pos ]))
343 return false;
```

```
344 pos --;
345 }
346 if (pos >= 0)
347 {
348 if (! parseChecksum(chunks[pos ]))
349 return false;
350 pos --;
351 }
352 if (pos >= 0)
353 {
354 if (! parseBlockSize (chunks[pos ]))
355 return false;
356 pos --;
357 }
358 if (pos >= 0)
359 {
360 if (! parseDescription (chunks[pos ]))
361 return false;
362 pos --;
363 }
364
365 m_groups[m_signalList.back (). m_filename]
366 .push_back(m_signalList.size () -1);
367
368 return true;
369 }
370
371 bool ECGData :: parseFileName(string chunk)
372 {
373 // analyzes the format the file name
374 m_signalList.back (). m_filename = chunk;
375
376 return true;
377 }
378
379 bool ECGData :: parseFormat (string chunk)
380 {
381 // analyzes the format of the data
382 sscanf_s(chunk.c_str(), "%ix%i:%i+%i"
383 , &m_signalList.back (). m_format
384 , &m_signalList.back (). m_samplesPerFrame
385 , &m_signalList.back (). m_skew
386 , &m_signalList.back (). m_byteOffset );
387
388 if (m_signalList.back (). m_format != 16) return false;
```

```
389
390 return true;
391 }
392
393 bool ECGData :: parseADCGain(string chunk)
394 {
395 char unit [2048];
396
397
398 // the format analyzes the AD increase
399 int r = sscanf_s(chunk.c_str(), "%f(%i)/%s"
400 , &m_signalList.back (). m_ADCGain
401 , &m_signalList.back (). m_baseline
402 , &unit
403 , sizeof(unit ));
404
405 if (r == 3) m_signalList.back (). m_units = unit;
406
407 return true;
408 }
409
410 bool ECGData :: parseADCResolution(string chunk)
411 {
412 // analyzes the format of the AD bit resolution
413 sscanf_s(chunk.c_str ()
414 , "%i"
415 , &m_signalList.back (). m_ADCResolution );
416
417 return true;
418 }
419
420 bool ECGData :: parseADCZero(string chunk)
421 {
422 // analyzes the format of the AD -zero value
423 sscanf_s(chunk.c_str ()
424 , "%i"
425 , &m_signalList.back (). m_ADCZero );
426
427 return true;
428 }
429
430 bool ECGData :: parseInitialValue(string chunk)
431 {
432 // analyzes the format of the initial value
433 scanf_s(chunk.c_str ()
```

```
434 , "%i"
435 , &m_signalList.back (). m_initialValue );
436
437 return true;
438 }
439
440 bool ECGData :: parseChecksum(string chunk)
441 {
442 // analyzes the format of the checksum
443 scanf_s(chunk.c_str ()
444 , "%i"
445 , &m_signalList.back (). m_checksum );
446
447 return true;
448 }
449
450 bool ECGData :: parseBlockSize(string chunk)
451 {
452 // analyzes the format of the block size
453 scanf_s(chunk.c_str ()
454 , "%i"
455 , &m_signalList.back (). m_blockSize );
456
457 return true;
458 }
459 bool ECGData :: parseDescription(string chunk)
460 {
461 // analyzes the format of the description
462 m_signalList.back (). m_description = chunk;
463
464 return true;
465 }
466
467 // opens and reads a data file
468 bool ECGData :: readDataFile(string filename)
469 {
470 string path = "";
471 string openFile = "";
472 fstream file;
473 int readSamples = 0;
474 char lsB = 0;
475 char msB = 0;
476 short sample = 0;
477
478 // Extract the path from the filename
```

```
479 path = filename.substr (0
480 , filename.find_last_of ( '\\ ')+1);
481
482 // new loop type
483 for (auto group : m_groups)
484 {
485 // As long as groups are still available
486 if (group.second.size () > 0)
487 {
488 // Open .dat file
489 file.open(path + group.first
490 , ios::in | ios:: binary );
491 if (! file.is_open ()) return false;
492
493 // Reading the signals from the file
494 while (! file.eof ())
495 {
496 for (int i = 0; i < group.second.size (); i++)
497 {
498 file.get(lsB);
499 file.get(msB);
500
501 // convert the two bytes into one
502 // two byte value
503 sample = (msB << 8) | lsB;
504
505 // Saving the value
506 m_signalList[group.second[i]]
507 .m_data.push_back(sample );
508 }
509 }
510
511 // Closing the file
512 file.close ();
513 }
514 }
515
516 return true;
517 }
```

Constructor, destructor and clear()

The task of the constructor is to initialize the member variables of the class, while the destructor is to free memory that may still be occupied by the object.

In this program, a function is also needed to ensure that the object of the class is returned to the start state if, for example, the loading of the data fails.

For this reason a function *clear( )* was written, which initializes the variables and clears the two arrays *m_signalList* and *m_groups*, i.e. restores the start configuration.

This function can be called from the constructor, the destructor, and from any other function to return the object to its start state.

loadECGData(string filename)

The *loadECGData* function implements the activity diagram in Fig. 12.10 at an abstract level. The activity nodes for loading and parsing the MIT header file have been offloaded to the *readHeaderFile* function. This function returns a *bool* value that the function can use to decide whether the load and parse were successful.

If the function was not successful, the return value is *false*. In this case, the variables of the class are set back to their initial state by the *clear( )* function and an error message is issued.

If the MIT header file is successfully read in and analyzed, an attempt is made to open and read out the associated files that contain the signal data. There is also an output in case of error.

If both actions are successful, the content of the collected information is printed from the comments of the MIT-HEader file and the function is terminated.

analyze() and storeResults()

The contents of these functions will not be discussed until the next two chapters, so only empty functions have been implemented so far.

readHeaderFile(string filename)

The function resets the variables of the class to the initial state first, in case a record has already been loaded.

After that, an attempt is made to open a file. For this purpose C++ offers the class *fstream*. This class is very closely related to the classes *istream* (*cin*) and *ostream* (*cout*), since it inherits from both classes.

In the function a variable *file* of the type *fstream* is created, with the help of which files can now be opened and read.

The *open* function of the *fstream* class allows a file to be opened. The first parameter is a text containing the full path and the file name. The second parameter contains information about how the file should be opened. In this case, the predefined value *ios::in* signals that it is about reading the contents of the file.

With the help of the function *is_open* it can now be checked whether the opening of the file was successful. In this case the data can be read out. Otherwise the function should abort with a *false* as return value.

Since it is not known how much data is contained in the file, a *while loop* is used. The termination condition checks whether the end of file [eof] has not yet been reached.

Within the loop, the *getline* function reads a complete line of data from the file opened with *file* and stores it in the variable *line*. For this purpose, data is read from the file until the character for an end of line ("\n") is found.

Finally, the *parseLine* function attempts to parse the contents of the line. If this is not successful, the file is closed by *file.close()* and terminated with *false* as return value.

parseLine(string line)

The task of the function *parseLine* is to distinguish how a read line from the MIT header file should be parsed.

First, it is checked whether the line starts with a hash (#) or is empty. The first case can be checked by using the *find* function provided by the class *string*. The return value corresponds either to the text position of the found character or to the predefined value *string::npos*. To check the second case, the length of the line is evaluated using the *length* function. If the length of the line is zero, the line is empty. In both cases the line is evaluated as a comment and appended to the *string m_info*.

Otherwise it is checked whether the *bool variable m_recordLine* contains the value *true*. This is only the case if a *record line* has already been read in. Consequently, the current line can only be a *signal specification line*, which is first preprocessed by the *getChunks* function and then passed to the corresponding function for analysis.

If the analysis of the line fails, the function terminates with the return value *false*.

The last option occurs if no *record line* has been read yet. In this case the line is also preprocessed by the *getChunks* function and passed to the corresponding analysis function. If this is successful, the value of the variable *m_recordLine* is set to *true* and the function exits with the return value *true*.

getChunks(string line)

A line within the *record line*, or the *signal specification line*, consists of several parameters, some of which are separated by special characters, such as a colon or brackets, or by spaces.

The *getChunks* function is supposed to break the line at the spaces and store the individual pieces (*chunks*) in a vector ⟨*string*⟩, that is, an array of texts.

To do this, the line is run through with a *while loop* as long as the line is not empty, i.e. as long as its length is greater than zero.

The *find_last_of* function is provided by the *string* class and finds the last occurrence of a character within a text. The return value is either the position of the character or the constant value *string::npos*.

If a space character could be found, then the text must be cut at this position. This task is performed by the *substr* function. The first parameter specifies the start position from which the text is to be taken and the second parameter specifies the number of characters to be cut. If this parameter is *string::npos*, the text is cut to the end.

The cut piece is attached to the end of the *chunks* array with the *push_back* function.

The line is then shortened to the appropriate position by storing the result of *substr(0, pos)* back into the line itself.

If no space is found, the last piece of text has been reached. This is appended to the array in full. The rest of the line can then be deleted. If this point is reached, the loop condition of the *while* loop will no longer apply in the next step, since the line length now corresponds to zero.

The function terminates by returning the array containing the collected pieces of text to the calling function.

parseRecordLine(vector<string> chunks)

In this function, the individual pieces of the *record line* are now analyzed in sequence. To do this, the start position is first placed at the end of the array, since the function *getChunks* splits the line from back to front.

The decomposition procedure strictly follows the syntax defined in Fig. 12.8 on page 224. All elements which are connected by special characters are located in a *chunk* and are extracted together in a subfunction.

The individual steps always follow the same pattern. First, it is checked whether the current position is greater than or equal to zero. Then a piece of text exists that can be analyzed. In this case, the piece of text is passed to the corresponding sub-function, which attempts to extract the contents. If this fails, the function exits with the return value *false*.

Otherwise, the position is decreased by the value one and it continues with the next section of the line.

This structure was chosen because the individual parts of the line, although largely optional, can only appear in a single order. Later parts of the line can only appear if the sections before them are available. For this reason, editing can also be aborted at any time without any problems.

At the end the function returns *true*.

parseRecordName(string chunk) to parseDate(string chunk)

After crunching and simplifying the data stage by stage, these functions now take care of the actual extraction of information.

For this purpose a C++ function called *sscanf_s is* used. The function can take an unlimited number of parameters, but needs at least the first two.

The first parameter is the text to be parsed and must be passed as an array of type *char.* The class *string* does not correspond to this data type, but can be converted to the required type by the function *c_str( )*.

The second parameter is a so-called format string and also an array of type *char.* However, it is usually passed directly as text in quotes.

This text describes how the first parameter is to be interpreted. Special characters mark information that is to be extracted from the text. The character *"%i"* stands for a value of type *int,* the character *"%f"* for a *double* and the character *"%s"* for a field of type *char.* Between these characters the expected formatting can be specified.

For example, *"%f/%f(%f)"* means to try to find a *double value* in the first parameter, followed by a slash and another *double*, then another *double* in parentheses.

The *sscanf_s* function now attempts to extract these values and store them in the appropriate order in its function parameters. These parameters must be passed as pointers. However, since the attributes of the class are not pointers, the variable names must be preceded by an ampersand.

An exception must be considered for text extraction with *"%s"*. Here one parameter must be an array of the type *char*, but then the size of the array must also be passed, so that the function *sscanf_s* does not accidentally exceed the array size. For this purpose the *C++* function *sizeof* is used for example in the function *parseRecordName*.

The return value of the function corresponds to the number of values found in the text. Since many parameters are optional in the *record line*, it can easily happen that values cannot be extracted. However, all of these parameters are to be ignored in this example, so the return value does not have to be evaluated.

parseSignalSpecificationLine(vector<string> chunks)

The structure of this function is basically identical to that of the function *parseRecordLine*. Here, too, the start position is set at the end of the array and the line is then analyzed piece by piece.

The analysis of the line strictly follows the structure shown in Fig. 12.9.

In addition, an object of the class *Signal* is created with each call of this function and appended to the end of the signal list *m_signalList*. Since the information of the *Signal Specification Line* is to be stored in the objects of the Signal class, a separate object must be created for each signal.

After a row has been successfully analyzed, an entry must also be stored in the group table *m_groups*. The data type *string* was defined as the index for the group table. The name of the file in which the signal is stored is to be stored in it. Behind this index there should be a list with index positions, which indicate which signals are in this file (see Table 12.1).

The current signal is always the last entry in the list. This can be obtained by *m_signalList.back()* and is of type *Signal*. In the *Signal* class, after the successful analysis of the *Signal Specification line,* the associated file name was stored in the member variable *m_filename*. This now serves as an index for the group table *m_groups[m_signalList".back()".m_filename]*.

The index of the current signal in the signal list is stored in the array at the position of the file name. Since this is always the last signal, the position can be calculated by *m_signalList.size()-1*.

The function ends with the return value *true,* provided that no error occurred while parsing the line.

parseFileName(string chunk) to parseDescription(string chunk)

The functions *parseFileName* to *parseDescription* parse the individual sections of the *Signal Specification line* using the function *sscan_f*, as described earlier.

The only difference now is that the variables in which the values are to be stored are located in an object of the *Signal* class. First, the correct object must be selected in order to be able to access the variable.

As already mentioned, the current object is always at the end of the signal list and can be queried by *m_signalList.back()*. Afterwards only the correct attribute must be completed with a dot.

readDataFile(string fileName)

In the last function the contents of the data files are read in. In this sample program only the already described *format 16* is supported. This simplifies the function considerably.

First, some auxiliary variables are created whose meaning is explained when they are needed.

In the parameter *filename* the path and the file name are passed, which were entered via the console. The same path is required for the data files, but the file name must be replaced.

For this reason, the *subst* function is used to extract the pure *pathname* by cutting out the beginning of the *filename* variable until the last appearance of the "\\" character.

What follows is a variation of the already familiar *for loop*, which has only been part of the *C++* language since 2011, and which greatly simplifies the use of the *map* data structure.

The structure is simple: First, a loop is started by the keyword *for*. In the loop header, however, the expression known from the *for loop* does not follow.

Instead, the keyword *auto* is used first in the brackets, followed by the name of a variable (*group* in this example). This is followed by a colon and the name of a data structure to be traversed. In this case, the *map m_groups*.

Within the loop, there will now be a different entry of the *map* in the variable *group* on each run. The variable has two member variables. The index of the *map* is stored in the variable *first*, in this case the file name as a *string,* and the value of the *map* is stored in the variable *second.* This was defined as a *vector ⟨int⟩* when the variable *m_groups* was declared.

Within the loop, it is first checked whether signals exist at all for the current group object. This is checked by querying the number of elements in the *vector* with *group.second.size()*. Only if elements exist, the opening and reading of the file makes sense.

Next, the file is opened. To do this, the file name must be appended to the already extracted file path. This is done by *path + group.first*. Since the data of the data files are not stored as texts, but binary, this must be explicitly specified when opening the file. This is done by connecting the two values *ios::in* and *ios::binary* with a binary OR.

The value *ios::in* corresponds to the numerical value 1, while *ios::binary* corresponds to the numerical value 32. Both numbers are necessarily powers of two. In binary, the two numbers correspond to the bit sequence 0000 0001, and 0010 0000, respectively. Thus, both numbers each consist of a bit sequence with only a single one. If these two numbers are combined with the binary OR, the result is the bit sequence 0010 0001. Based on this bit sequence, the *open* function can recognize which options were selected for opening the file. Values that are linked and processed in this way are called *flags*.

If the file cannot be opened, the function terminates with the return value *false*.

To read out the data, a *while loop* is used again, which runs through the file to the end.

The data within the file is arranged so that the first value for the first signal is followed by the first value for the second signal. The order of the signals is identical to the order in the MIT header file. If all signals have been passed through once, the second value for each signal follows, and so on.

In the implementation, a *for loop* is used to step through the list of signal indices from the beginning to the end.

Each value in the file is stored in two bytes, the first of which contains the bits with lower values and the second the bits with higher values. This ordering of the bytes is called *little-endian*. The reverse order is called *big-endian*.

Unfortunately, the already known operation $\rangle\rangle$ cannot be used for reading the data, because it would skip supposed spaces. However, since the file was saved in binary, there are no spaces.

Instead, the *get* function must be used, which reads exactly one byte from the file. After the byte with the *least significant* bits *(lsB)* has been read, the byte with the *most significant* bits *(msB)* follows.

Both bytes must now be combined to form a *short*. This is done by shifting the *msB* by 8 bits to the left by *(msB $\langle\langle$ 8)* and linking it to the *lsB* with a binary OR.

The byte sequence 17 *FE*, which consists of the two bytes 0001 0111 and 1111 1110 in binary form, is to serve as an example. The second of the two bytes is shifted 8 bits to the left, resulting in the bit sequence 1111 1110 0000 0000. If this bit sequence is linked to the first byte by a binary OR, the result is the bit sequence 1111 1110 0001 0111. This corresponds to the number -489. In the example data used, this corresponds exactly to the control value that can be taken from the *header file*.

Next, the value must be stored in the correct object of the *Signal* class. Since the loop variable *i* passes through the list of signal indices, the correct index is located at the position *group.second[i]*. If this position is queried from the field *m_signalList*, the object searched for results. After that, the read value only has to be appended to the array of the signal data by *.m_data.push_back(sample)*.

Finally, the open file is closed and the return value *true* is returned to the calling function.

If everything went successfully, all data is now within the data structures of the program.

Of course, this implementation does not yet meet the requirements of a professional software. Many special and error cases were ignored. However, this was also not the goal.

With the implementation presented here a relatively simple possibility should be shown to read in data of the MIT format.

12.3   Data Analysis

Now that the ECG data can be read into the software, thought can be given to the processing of the data. Here, too, the program should be limited to the essentials. When processing data, more or less complex mathematical procedures are often required. These procedures must be understood in order to be able to implement them in a program. For this reason, the applied procedure of the *Fast Fourier Transformation* is explained in detail in this chapter.

Since the read-in data are very noisy, a possibility should be found to highlight the interesting signals more clearly. One possibility for this is the so-called *Fourier transformation* (FT). In the FT, a continuous function *f(t)* represented over time *t* is transformed by integration into a function $F(\omega)$ dependent on the angular frequency ω. With the help of this transformation, the frequencies that make up the signal can be determined, so that only the frequencies relevant to the ECG remain through a filter operation. Afterwards, an *inverse Fourier transform* (IFT) could be used to recover the now filtered signal.

An easy introduction to this topic is provided by the derivation of the *Fourier series* presented in Papula (2015).

However, the read-in data is not in the form of a continuous function. Instead, values $h(t_k) = h_k$ were recorded at certain discrete points in time t_k and stored in a file. However, the FT can also be used for this type of data by means of a transformation. It is then called a discrete *Fourier transform* (DFT). Equation 12.1 shows the central formula of the DFT.

$$H_n = \sum_{N-1}^{k=0} h_k e^{-\frac{2\pi k n}{N}i}$$

(12.1)

The values *n* and *k* are integer values with *n, k* ∈ {0, 1, 2, ..., *N* − 1} and correspond to the indices of the respective time points t_k, respectively the frequencies f_n. The time t_k can be calculated by the formula $t_k = \frac{k}{R}$, where $R \in N$ stands for the sampling rate. The frequency results from the formula $f_n = n\frac{R}{N}$, with $N \in N$ recorded values. A very nice representation of the DFT can be found in Smith (1997).

The result of the DFT are complex values $H_n = a + ib$, which contain the amplitude and the phase angle for the respective frequency. The amplitude results from the magnitude of the complex number to ⊦ $H_n\# = \sqrt{a^2 + b^2}$, while the phase angle can be calculated in C++ via the function *atan2(b, a)*.

The inverse DFT (IDFT), which can be used to transform the individual frequencies back into the time domain, is shown in Eq. 12.2. The formula differs from the normal DFT equation only by one sign and one factor.

$$h_k = \frac{1}{N} \sum_{N-1}^{n=0} H_n e^{\frac{2\pi k n}{N}i}$$

(12.2)

In principle, formulas 12.1 and 12.2 can be used directly for an implementation of the DFT, or IDFT. However, in computer science, an implementation has gained acceptance that is traced back to an article by *James Cooley* and *John Tuckey* (Cooley and Tukey 1965). This implementation uses various properties of the complex numbers to speed up the computation of the DFT. For this reason, the method is called *Fast Fourier Transform* (FFT). A number of FFT variants exist, the simplest and most intuitive of which will be implemented here.

To speed up the DFT, the sum in formula 12.1 is first split into two summands, as shown in Eq. 12.3, adding up the even and odd parts of the formula respectively. A closer look then reveals that the exponent of the e-function can first be multiplied out and then decomposed to give a product of two e-functions. The exponent of the first e-function is then only dependent on n and can be drawn before the sum. After the decomposition, the second e-function is identical to the one in the sum of the even parts.

$$H_n = \sum_{N-1}^{k=0} h_k e^{-\frac{2\pi kn}{N}i}$$

$$= \sum_{N/2-1}^{k=0} h_{(2k)} e^{-\frac{2\pi(2k)n}{N}i} + \sum_{N/2-1}^{k=0} h_{(2k+1)} e^{-\frac{2\pi(2k+1)n}{N}i}$$

$$= \sum_{N/2-1}^{k=0} h_{(2k)} e^{-\frac{2\pi kn}{(N/2)}i} + \sum_{N/2-1}^{k=0} h_{(2k+1)} e^{-\frac{2\pi n}{N}i} e^{-\frac{2\pi kn}{(N/2)}i}$$

$$= \sum_{N/2-1}^{k=0} h_{(2k)} e^{-\frac{2\pi kn}{(N/2)}i} + \underbrace{e^{-\frac{2\pi n}{N}i}}_{g^n} \sum_{N/2-1}^{k=0} h_{(2k+1)} e^{-\frac{2\pi kn}{(N/2)}i}$$

$$= H_n^{gerade} + g^n H_n^{ungerade}, \text{mit } g^n = e^{-\frac{2\pi n}{N}i} \tag{12.3}$$

If a measurement series with N elements is to be transferred into the frequency space by the FFT, two data series must be generated from the data series, each consisting only of the even or odd elements of the original data series. These are subdivided again, and again, and again, until the data series consist of only a single element. In this case, k is always 0 and the e-function takes the value 1. Thus, the Fourier transform of a one-element data series is the element itself. To then form the Fourier transforms for longer data series, only the sum of the even element with the product of g^n and the odd element must be formed according to formula 12.3.

The principle according to which this algorithm works is called *Divide & Conquer*. A problem is divided into smaller problems until the solution of the smallest subproblem is quite simple and the more complex solutions can be assembled from the parts. The implementation of this concrete algorithm can be realized very easily and directly with a recursive function.

Basically, the procedure uses two values from the original data series to generate a new, complex value. The assumption is obvious that only half the size of the data series can be generated in this way, because how should the information content be doubled by the procedure? Eq. 12.4 examines the case in which the value n is in the upper half of the data series, i.e. shifted by the value $\dfrac{N}{2}$.

$$H_{n+\frac{N}{2}} = \sum_{N/2-1}^{k=0} h_{(2k)} e^{-\frac{2\pi k\left(n+\frac{N}{2}\right)}{(N/2)}i} + e^{-\frac{2\pi\left(n+\frac{N}{2}\right)}{N}i} \sum_{N/2-1}^{k=0} h_{(2k+1)} e^{-\frac{2\pi k\left(n+\frac{N}{2}\right)}{(N/2)}i}$$

$$= \sum_{N/2-1}^{k=0} h_{(2k)} e^{-\frac{2\pi kn}{(N/2)}i} \underbrace{e^{2\pi ki}}_{1} + e^{-\frac{2\pi n}{N}i} \underbrace{e^{\pi i}}_{-1} \sum_{N/2-1}^{k=0} h_{(2k+1)} e^{-\frac{2\pi kn}{(N/2)}i} \underbrace{e^{2\pi ki}}_{1}$$

$$= \sum_{N/2-1}^{k=0} h_{(2k)} e^{-\frac{2\pi kn}{(N/2)}i} - \underbrace{e^{-\frac{2\pi n}{N}i}}_{g^n} \sum_{N/2-1}^{k=0} h_{(2k+1)} e^{-\frac{2\pi kn}{(N/2)}i}$$

$$= H_n^{gerade} - g^n H_n^{ungerade}, \text{mit } g^n = e^{-\frac{2\pi n}{N}i} \tag{12.4}$$

In the equation, the value $n+\dfrac{N}{2}$ is inserted instead of n, otherwise the equation remains unchanged. Again, the exponents of the e-functions can be multiplied out and decomposed, so that the original e-functions reappear. The exponents of the split-off e-functions then consist only of constant parts, or the value k, which, however, is always multiplied by 2π, so that the result of the e-function is always a constant 1 or -1.

If the function proportions are then combined as in Eq. 12.3, it can be seen that the second half of the data can be calculated quite easily from the first half by simply converting the sum to a difference.

For the procedure to work, the size of the data series cannot, of course, be chosen completely freely, because the repeated halving of the data series must always lead to one-element data series in the end. The size N of the data series must therefore always be a power of two. In most cases, however, this can be ensured without any problems, since it can already be taken into account when planning an experiment.

In the articles by Goovaerts et al. (1976), by Thakor et al. (1983) or in the description of the Pan-Tompkins algorithm for the detection of the QRS complex (Pan and Tompkins 1985), it is indicated that the QRS complex can be found in a frequency range of 5–15 Hz. After the read-in data set has been decomposed into its frequencies using the FFT, the corresponding frequency band can be filtered out using a band-pass filter.

Finally, the data must be back-transformed using inverse FFT (IFFT) to represent the QRS complex.

12.3.1 Extension of the Software Architecture

Although analyzing data is a very challenging task from a mathematical point of view, developing a software architecture for the example program in this book is, however, very straightforward. Figure 12.12 shows an activity diagram that describes the processes within the *Analyze ECG Data* activity node.

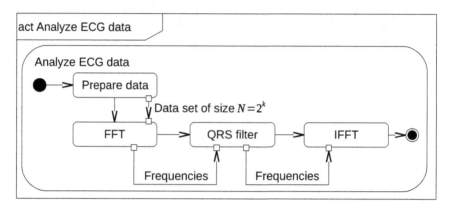

Fig. 12.12 Activity diagram for the analysis of ECG data

After the node has been activated, the read-in data must first be prepared to enable the application of the *Fast Fourier Transformation*. For this purpose, the size of the read-in data set must be determined and the next smaller power of two must be found. Then, a portion of the original data must be transformed into a new data set whose size corresponds to a power of two and which is either smaller than or equal to the original size.

Of course, it would also be possible here to allow the user to select between different window sizes and an exact positioning of the data window, but the software should not become unnecessarily complicated. A check whether data have already been read in is not necessary, since in the worst case a data set of length 0 would be transformed.

After the data has been processed, the *Fast Fourier Transform* is performed on the new data set. This transforms the time signal into its frequencies. Now a band-pass filter can be used, which fades out all frequencies outside the range of 5–15 Hz.

Finally, the frequencies must be converted back into a temporal signal by the IFFT, in which the QRS complex should now be more prominent. This ends the activity of the node.

The class diagram must be extended with some additional functions, as shown in Fig. 12.13.

First of all, the class *ECGData* has to be extended by two operations, which realize the normal and the inverse *Fast Fourier Transformation*. In addition, a function for the QRS filter is to be implemented. All three functions require an array of complex numbers on which they can perform their operations.

The individual steps of the calculation are to be saved in the *Signal* class so that the data can be printed later. Two new attributes are added for this purpose. The array *m_freq* is to store the frequencies after the FFT, while the array *m_processedData* is to save the result of the entire conversion. Both arrays must be able to store complex numbers.

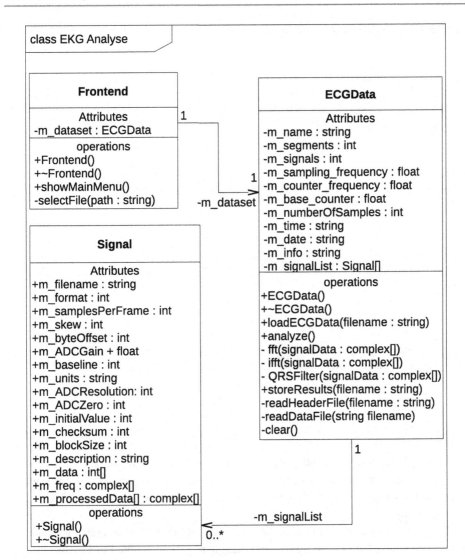

Fig. 12.13   Second development of the class diagram for ECG analysis

12.3.2 Implementation of the *Fast Fourier Transformation*

The changes in the *Signal* class are easy to implement because only two new arrays need
to be added. Listing 12.7 shows the changes in the *header file of* the *Signal* class.

Listing 12.7 The Signal class (Signal.h)

```
1 // ...
2 #include <string >
3 #include <vector >
4 #include <complex >
5 using namespace std;
6
7 class Signal
8 {
9 public:
10 Signal ();
11 ~Signal ();
12
13 // Signal Specification Line Data
14 // ...
15
16 // Signal data
17 vector <int> m_data;
18
19 // Frequency spectrum
20 vector <complex <double >> m_freq;
21
22 // Processed data
23 vector <complex <double >> m_processedData;
24 };
```

To be able to store the results, arrays of complex numbers have to be created. For the arrays, the data type *vector* has already been introduced, which is provided by the standard template library. In addition, the data type *complex* is now to be used, which can store complex numbers.

The declaration of the variable *m_freq* can now be interpreted as creating a complex number ***complex⟨double⟩*** whose real and imaginary parts are stored in variables of type *double*. The declaration *vector ⟨complex ⟨double⟩⟩ m_freq;* creates an array of such complex numbers and gives the array the name *m_freq*.

The *.cpp file of* the *Signal* class is shown in Listing 12.8.

Listing 12.8 The Signal class (Signal.cpp)

```
1 // ...
2
3 Signal ::~ Signal ()
4 {
5 m_data.clear ();
```

```
6 m_freq.clear ();
7 m_processedData.clear ();
8 }
```

The changes are limited to the destructor of the *Signal* class, in which the new data arrays are now also cleared before the object is deleted.

A number of new function declarations must first be created in the *ECGData* class. Listing 12.9 shows the changes in the class *header* file.

Listing 12.9 The ECGData class (ECGData.h)

```
 1 #include "Signal.h"
 2 #include <string >
 3 #include <vector >
 4 #include <map >
 5 #include <complex >
 6
 7 using namespace std;
 8
 9 class ECGData
10 {
11 public:
12 // ...
13
14 void analyze ();
15
16 // ...
17 private:
18 // ...
19
20 // Fast Fourier Transformation
21 void fft(vector <complex <double >> &signalData );
22 void ifft(vector <complex <double >> &signalData );
23 void ifft_r(vector <complex <double >> &signalData );
24
25 // Filter
26 void QRSFilter(vector <complex <double >> &signalData );
27
28 // ...
29 };
```

Essentially, the functions are generated that were already planned in the software architecture. However, the IFFT was split into two functions, since their calculation must be implemented recursively. This is done in the function *ifft_r*. Finally, however, the result must be multiplied by a factor. For this purpose, the function *ifft* was implemented, which is also used for calling the IFFT.

All functions use references to arrays of complex numbers as transfer parameters, as they are already used in the *Signal* class. The references are necessary so that the changes within the functions also remain outside the function.

The changes in the *.cpp* file are again very extensive. The presentation in Listing 12.10 is therefore limited to the newly added functions. An explanation of the individual functions follows after the program.

Listing 12.10 The ECGData class (ECGData.cpp)

```
 1 #include "stdafx.h"
 2 #include "ECGData.h"
 3 #include <iostream >
 4 #include <fstream >
 5
 6 using namespace std;
 7
 8 const double M_PI = 3.14159265358979323846;
 9
10 // ...
11
12 void ECGData :: analyze ()
13 {
14 // Run and conversion of all loaded
15 // Signals
16 for (int i = 0; i < m_signalList.size (); i++)
17 {
18 // Calculation of the next lower power of two
19 int twoExp = log2 (m_signalList[i]. m_data.size ());
20 int maxIndex = pow(2, twoExp );
21
22 // Create an auxiliary array for the data
23 vector <complex <double >> signalData;
24
25 // Copying the data series up to the calculated
26 // Power of two into the auxiliary array
27 for (int j = 0; j < maxIndex; j++)
28 {
29 signalData.push_back(
30 (double)m_signalList[i]. m_data[j] /
31 m_signalList[i]. m_ADCGain );
32 }
33
```

```
34 // Carrying out the FFT
35 fft(signalData );
36
37 // Saving the result in the data series
38 m_signalList[i]. m_freq = signalData;
39
40 // Applying the QRS filter
41 QRSFilter(signalData );
42
43 // Back transformation of the data
44 ifft(signalData );
45
46 // Saving the result in the data series
47 m_signalList[i]. m_processedData = signalData;
48 }
49
50 cout << "Analysis completed successfully !"
51 << endl << endl;
52
53 cin.get ();
54 }
55
56 void ECGData ::fft(vector <complex <double >> &signalData)
57 {
58 // Termination for one-element data series
59 if (signalData.size () == 1) return;
60
61 // Auxiliary arrays for even and odd
62 // Elements
63 vector <complex <double >> odd;
64 vector <complex <double >> even;
65
66 // Auxiliary variable for half the array size
67 int N = signalData.size () / 2;
68
69 // Copying the even and odd elements
70 // in the auxiliary arrays
71 for (int i = 0; i < N; i++)
72 {
73 even.push_back(signalData [2 * i]);
74 odd.push_back(signalData [2 * i + 1]);
75 }
76
77 // Recursive call of the FFT
78 fft(even );
```

```
79 fft(odd);
80
81 // Combining the results
82 for (int i = 0; i < N; i++)
83 {
84 complex <double> g = polar (( double )1,
85 (double)-M_PI * i / N);
86 signalData[i] = even[i] + g * odd[i];
87 signalData[i + N] = even[i] - g * odd[i];
88 }
89 }
90
91 void ECGData :: ifft(vector <complex <double >> &signalData)
92 {
93 // recursive calculation of the IFFT
94 ifft_r(signalData );
95
96 // Multiplication of the result by the
97 // Prefactor 1 / N
98 for (int i = 0; i < signalData.size (); i++)
99 {
100 signalData[i] /= signalData.size ();
101 }
102 }
103
104 void ECGData :: ifft_r(vector <complex <double >> &signalData)
105 {
106 // Termination for one-element data series
107 if (signalData.size () == 1) return;
108
109 // Auxiliary arrays for even and odd
110 // Elements
111 vector <complex <double >> odd;
112 vector <complex <double >> even;
113
114 // Auxiliary variable for half the array size
115 int N = signalData.size () / 2;
116
117 // Copying the even and odd elements
118 // to the auxiliary arrays
119 for (int i = 0; i < N; i++)
120 {
121 even.push_back(signalData [2 * i]);
122 odd.push_back(signalData [2 * i + 1]);
123 }
```

```
124
125 // Recursive call of the IFFT
126 ifft_r(even );
127 ifft_r(odd);
128
129 // Combining the results
130 for (int i = 0; i < N; i++)
131 {
132 complex <double> g = polar (( double )1,
133 (double)M_PI * i / N);
134 signalData[i] = even[i] + g * odd[i];
135 signalData[i + N] = even[i] - g * odd[i];
136 }
137 }
138
139 void ECGData :: QRSFilter
140 (vector <complex <double >> &signalData)
141 {
142 // Auxiliary variable for half the array size
143 int N = signalData.size () / 2;
144 // Calculation of the index for 5Hz
145 int i_min = 5 * signalData.size () /
146 m_sampling_frequency ;
147 // Calculation of the index for 15 Hz
148 int i_max = 15 * signalData.size () /
149 m_sampling_frequency ;
150
151 // Blanking of all frequencies outside the
152 // desired spectrum
153 for (int i = 0; i < N; i++)
154 {
155 if (i < i_min || i > i_max)
156 {
157 signalData[i] = 0;
158 signalData[i + N] = 0;
159 }
160 }
161 }
162
163 // ...
```

analyze()

In the *analyze function,* the list of all loaded signals must be traversed in order to perform the analysis on all data sets.

To do this, the power of two must first be found that represents the size of the data array as well as possible. From this, the index can be calculated up to which the data can be evaluated.

All elements of the record up to this index are stored in an auxiliary array of complex numbers, which is required for further processing.

Since the stored data is raw data resulting from digitization by the AD converter, it must still be divided by the AD increment to restore the value of the original analog signal.

Now the FFT can be performed to calculate the frequencies of the signal. These are to be saved for later output in the *Signal* class.

The *QRSFilter* function then restricts the frequencies to the range of 5–15 Hz and the *ifft* function converts them back into a time signal. This result is also to be saved for later output.

fft(vector <complex double >> & signalData)
The function fft is based on an implementation of Rosettacode (2019) and puts into practice the considerations from Eqs. 12.3 and 12.4.

If the signal is divided further and further into even and odd elements, one-element data series will eventually result. In this case, the calculation may terminate because the result corresponds to the one element.

Otherwise, two auxiliary arrays must be created in which the even and odd elements can be saved. This is done using a *for loop* that splits the elements into the auxiliary arrays according to their position.

The FFT can then be applied recursively to the two subarrays. If the arrays still consist of several elements, they are now split. Otherwise, the one element is returned as the result.

Finally, the results of the recursive calls are combined into a complex number. For this purpose, the formula from Eq. 12.3 is used for the first half of the data, while at the same time the results of the second half of the data are calculated using Eq. 12.4.

ifft(vector <complex <double>> & signalData)
To calculate the inverse FFT, a very similar procedure must be gone through as with the normal FFT. However, the result must be multiplied once by the factor $\frac{1}{N}$.

For this reason, the procedure was distributed over two functions. In the function *ifft*, the function *ifft_r* is called first, which (similar to the normal FFT) initially takes over the recursive calculation of the data.

Finally, each element of the data series is divided by the value N within a loop.

ifft_r(vector <complex <double >> & signalData)
The calculation of the recursive part of the inverse FFT is identical to that in the function *fft*, except for one sign.

QRSFilter(vector <complex <double >> & signalData)

To limit the frequency spectrum, different methods can of course be used. In professional software, it would make sense to implement different tools individually in order to then combine them into more complex processes.

In this example, a separate function could be written for a high and low pass filter. The QRS filter would then only have to execute both in succession to achieve the desired band-pass effect.

In this example, however, a specialized function is to be developed.

For this purpose, the indices at which the frequencies 5 Hz and 15 Hz are located are calculated first.

The FFT assumes in its calculation that the transferred data series comprises exactly one second. In this case, exactly the value for the frequency 1 Hz would also be found at index position 1.

If the recorded signal is longer than one second, the index position must be divided by the number of recorded seconds to obtain the frequency. The number of seconds results from the total number of data (*signalData.size()*) divided by the number of recordings per second (*m_sampling_frequency*).

So the equation can be rearranged to calculate the index position from the desired frequency.

$$\omega = i \cdot \frac{m\_sampling\_frequency}{signalData.size(\ )} \Leftrightarrow i = \omega \cdot \frac{signalData.size(\ )}{m\_sampling\_frequency} \qquad (12.5)$$

Finally, only all values outside of these index positions must be set to 0. Note that this process must be applied to the upper and lower half of the data as before.

12.4 Exporting the Results

When exporting the results, it is important to choose an export format that can be processed as easily as possible. The *.csv format* is a good choice here. The abbreviation stands for *Comma-Separated-Values* and describes a data format in which different values can be sorted in columns by separating them with commas.

In fact, however, the separation of values often takes place by other characters as well, and most spreadsheet programs allow you to freely choose the separator when importing such a file. Common separators are semicolons or tabs.

During import and data analysis, three different data sets were generated for each signal to be exported together.

- The raw data from the measurement, which must be converted to analog data before export.
- The frequencies calculated by the *Fast Fourier Transform.*

- The back-transformed data after the QRS filter was applied.

All three records are in arrays.

Since most spreadsheet programs assume that long data sets are in columns, the data should be arranged accordingly.

12.4.1 Extension of the Software Architecture

In order to be able to export the data, a file must be opened, as was already the case with the data import. This can lead to various errors, even if the file to be written to is newly created.

For example, what happens if the file already exists, no write permissions have been granted in the destination folder, and so on.

Figure 12.14 shows the activity diagram for the *Export Results* activity node.

First, an attempt is made to open the *.csv file*. If this fails, an error output must occur and the processing of the activity must be terminated.

Otherwise, you can continue with the export of the data.

The class diagram does not need to be changed because the *storeResults* function was planned into the architecture from the beginning.

12.4.2 Implementation of the Export Function

Listing 12.11 shows the adjustments made to the *header* file to export the data.

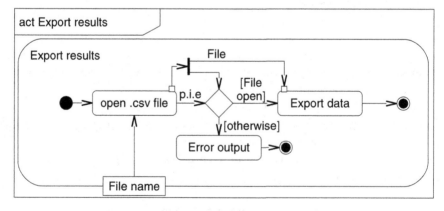

Fig. 12.14 Activity diagram for exporting the results

Listing 12.11 The ECGData class (ECGData.h)

```
1 #include "Signal.h"
2 #include <string >
3 #include <vector >
4 #include <map >
5 #include <complex >
6
7 using namespace std;
8
9 class ECGData
10 {
11 public:
12 // ...
13
14 void storeResults(string filename );
15
16 // ...
17 private:
18 // ...
19
20 // Output
21 void printHeadline(fstream &file ,
22 string unit ,
23 string postfix );
24
25 // ...
26 };
```

In principle, the export can be handled in a single function. However, in order to improve clarity, the function *printHeadline* has been added, which is intended to allow standardized headlines to be printed for the table.

Listing 12.12 shows the implementation of the export function in the *.cpp* file of the *ECGData* class. The explanation of the functions follows the program text.

Listing 12.12 The ECGData class (ECGData.cpp)

```
1 #include "stdafx.h"
2 #include "ECGData.h"
3 #include <iostream >
4 #include <fstream >
5
6 using namespace std;
7
```

```
 8 const double M_PI = 3.14159265358979323846;
 9
10 // ...
11
12 void ECGData :: storeResults(string filename)
13 {
14 // Variable declaration
15 fstream file;
16
17 // Opening the file for export
18 file.open(filename , ios::out);
19
20 // If the file could be opened
21 if (file.is_open ())
22 {
23 // Output of the headings for the
24 // three data blocks
25 printHeadline(file , "t (s)", "");
26 printHeadline(file , "f (HZ)", "_f ");
27 printHeadline(file , "t (s)", "_p ");
28
29 file << endl;
30
31 // Run through all read-in data records
32 for (int i = 0; i < m_numberOfSamples ; i++)
33 {
34 // Output of the time in seconds
35 file << double(i) / m_sampling_frequency << ";";
36
37 // Output of the read-in data for
38 // any signal
39 for (int j = 0; j < m_signalList.size (); j++)
40 {
41 // Confirmation prompt , so that the length of the
42 // data record can never be exceeded
43 //
44 if (i < m_signalList[j]. m_data.size ())
45 file << double(m_signalList[j]. m_data[i]) /
46 m_signalList[j]. m_ADCGain << ";";
47 else
48 file << ";";
49 }
50
51 // Frequency output
52 file << double(i) * m_sampling_frequency /
```

```
53 m_numberOfSamples << ";";
54
55 // Output of the calculated frequency spectra
56 // for each signal
57 for (int j = 0; j < m_signalList.size (); j++)
58 {
59 // Confirmation prompt , so that the length of the
60 // half of the data set can never be exceeded
61 //
62 // Data in the second half
63 // do not bring any new insights
64 if (i < m_signalList[j]. m_freq.size () / 2)
65 file << double(abs(m_signalList[j]. m_freq[i]))
66 << ";";
67 else
68 file << ";";
69 }
70
71 // Output of the time in seconds
72 file << double(i) / m_sampling_frequency << ";";
73
74 // Output of the retransformed data
75 // for each signal
76 for (int j = 0; j < m_signalList.size (); j++)
77 {
78 // Confirmation prompt , so that the length of the
79 // data record can never be exceeded
80 //
81 if (i < m_signalList[j]. m_processedData.size ())
82 file << double(m_signalList[j].
83 m_processedData[i].real ())
84 << ";";
85 else
86 file << ";";
87 }
88
89 file << endl;
90 }
91
92 // Closing the open file
93 file.close ();
94
95 cout << "Data export completed successfully !"
96 << endl << endl;
97
```

```
98 cin.get ();
99 }
100 else
101 {
102 // Error message
103 cout << "Data export failed !"
104 << endl << endl;
105
106 cin.get ();
107 }
108 }
109
110 void ECGData :: printHeadline(fstream &file ,
111 string unit ,
112 string postfix)
113 {
114 // Output of the unit for the x-axis
115 file << unit << ";";
116
117 // Output of the headings of the individual signals
118 for (int j = 0; j < m_signalList.size (); j++)
119 {
120 file << m_signalList[j]. m_description
121 << postfix << ";";
122 }
123 }
124
125 // ...
```

storeResults(string filename)

The function first tries to open the file whose name was entered by the user. If this is successful, the various data records are exported. Otherwise, the function issues an error message and cancels further processing.

The function must always process a row completely before jumping to the next row, since a return to a previous row would be extremely time-consuming. Since the data is to be in the columns, this means that the first element for each data record must be printed first, then the second, and so on.

The headings are to be found in the first row. For this reason, a series of headings is created for each data record. The first data set consists of the read-in signal data for each signal. The unit of the x-axis is the time in seconds. The second data set consists of the calculated frequencies for each signal. The unit of the x-axis is the frequency in hertz. In addition, each heading should have the suffix _f to indicate that it is the frequencies. The third data set is the back-transformed signals after the QRS filter. The unit of the x-axis is again the time in seconds. All headings should have the addition _p (for *processed*).

All possible indices are now run through in an outer loop. The maximum value *m_numberOfSamples* was read in from the MIT file.

First, the x-coordinate of the first data set is calculated. The time in seconds results from the current index divided by the number of signals per second.

After that, the *i-th* element of the signal data is printed for each signal. These are still divided by the AD increment to obtain the values of the analog signal. A safety query prevents access beyond the end of the array, should an array be unexpectedly shorter than the others.

The x-coordinates of the second data set are the frequencies. These result from the current index divided by the length of the recording in seconds. This is calculated by dividing the number of signals by the number of signals per second.

For each signal, the data is now printed. Since the frequencies are complex numbers, the real and imaginary parts would have to be printed. Instead, the *abs* function calculates the magnitude of the complex number. This corresponds to the amplitude of the oscillation at the corresponding frequency.

For this data set, only half of the data is printed because the second half has symmetry to the first half.

For the third data set, the time in seconds is calculated again to determine the values of the x-axis.

Here, too, the data are printed individually for each signal. After the reverse transformation, the imaginary parts of all numbers should have the value 0. For this reason, only the real part of the complex numbers is printed by the function .*real()*.

Individual data are separated from each other by semicolons. This ensures that later each value ends up in its own column. The end of a line is inserted with *endl*.

After all data has been printed, the file is closed and a success message is printed.

printHeadline(fstream&file, string unit, string postfix)
The output of the headings always follows the same pattern, therefore this functionality has been outsourced to a separate function.

The parameters of the function describe the file to be written to, the unit of the values of the x-axis and a supplementary text to be appended to the normal name of the heading.

First the unit of the x-axis is printed and separated from the other outputs by a semicolon. Then the descriptions of each signal read in are printed as a heading and provided with the supplementary text. The separation of the individual headings is again done by semicolons.

No *endl* is printed in the function, since the function should not determine the end of a line. This allows the output of the headings for the three data sets in one line.

12.4.3 Presentation of the Results

Now that all requirements for the sample program have been met, a look at the finished system is to be taken. In principle, *compiling* should already be done frequently during the development of a program in order to detect typing errors in the program as early as possible. In addition, the area to be searched is significantly smaller if the program is compiled frequently.

But after the program has been compiled, it does not necessarily do what it is supposed to do. Extensive tests of the functionality should therefore always be planned. It is best to plan standardized tests that cover every part of the program if possible. Since the program developed here is very simple, only each part of the program needs to be called once to cover all parts of the program. In addition, however, at least one attempt should be made at each input to willfully enter something incorrect.

The program reports with the main menu and a prompt:

```
Welcome to the ECG analysis program!
The following options are available to you:
1. Load ECG data
2. Analyze ECG data
3. Export results
4. Exit program
Please make your selection:
```

A small bug has already been built into the prompt so you can find and fix something.

After the program has been tested, we will have a look at the results. Figure 12.15 shows an ECG signal downloaded from the *Physionet* website (Goldberger et al. 2000) from the Physikalisch-Technische Bundesanstalt (PTB) database (Bousseljot et al. 1995). The figure shows an I-extremity derivative in a 12-lead ECG according to Eindhoven.

As can be clearly seen, the signal is very noisy and the individual phases of the QRS complex can only be recognised with difficulty, even with the human eye. For this reason,

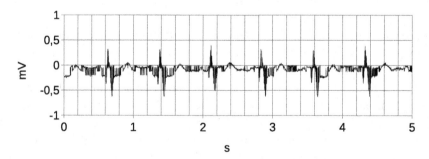

Fig. 12.15 Representation of a noisy I derivative to Eindhoven

the frequency spectrum, which is shown in Fig. 12.16, will first be calculated with the aid of the program developed here.

The higher a value is displayed on the y-axis, the greater is the amplitude of the oscillation with the corresponding frequency. The sum of all oscillations results in the original signal. It can be clearly seen that the oscillations with the greatest amplitude are to be found in the range up to approx. 15 Hz. Part of this range is cut out by the QRS filter and the sum of the remaining oscillations gives a new signal, which is freed from higher frequencies with lower amplitude. The result is shown in Fig. 12.17.

Overall, the amplitude of the oscillation has decreased, but the individual heartbeats with the QRS complexes resulting in the signal can now be detected much more easily.

If this were a professional software, the project would of course be far from finished. More analyses can be implemented and a graphical interface that simplifies the operation would certainly also be an advantage. Computer science is a very broad field with many specializations and with this book you have only taken the first step into this world.

I hope you have fun and good luck exploring the world of computer science!

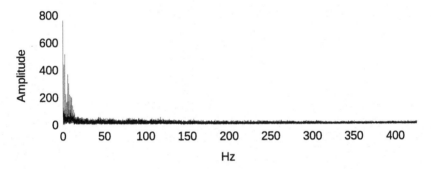

Fig. 12.16 Plot of the amplitude of the frequency spectrum of the I derivative

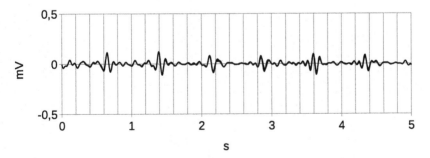

Fig. 12.17 Representation of the I-derivation after application of the QRS filter

Solutions

Chapter 5

Variable Definition

In order to create a variable in a C++ program, at least two pieces of information are required.

- First, the data type must be known so that it is determined how much memory is required for the variable and how the data it contains is to be interpreted.
- Secondly, a unique name must be defined for the variable so that the *compiler* can recognize when the variable is to be accessed.

Memory
The individual variable types occupy the following amount of memory:

(a) A *char* occupies exactly 1 byte of memory.
(b) For a *short,* the size depends on the processor used, but is usually in the range of 2 to 8 bytes and is less than or equal to an *int*.
(c) The memory usage of a *float* s is 4 bytes.
(d) The data type *int* behaves like the data type *short*. Depending on the processor, the data type has a size of 2 to 8 bytes, and is greater than or equal to *short*.

© The Author(s), under exclusive license to Springer Fachmedien Wiesbaden GmbH, part of Springer Nature 2023
B. Tolg, *Computer science to the Point,*
https://doi.org/10.1007/978-3-658-38443-2

(e) A variable of type *void* cannot be created because the variable type does not occupy any memory.

Typecast

A *typecast* makes a variable of type *A* look like a variable of type *B for* the duration of an operation. In *C++*, a *typecast is* created by writing the new variable type in round brackets before the value to be interpreted. An example looks like this:

```
/ ...
double value = 3.5;
cout << (int)value << endl;
// ...
```

Enumerations

Constant values are often required in programs to code different states. It quickly becomes confusing if only numbers are used, as it is easy to forget what a number stands for. If the numbers are stored in variables, the programmer is still responsible for ensuring that there are no duplicates in the numbers.

An enumeration allows the programmer to assign names to constant values and let the *C++* language manage the associated numbers. This increases the readability of programs, which are also easier to maintain as a result.

Variable Definition in C++

A *declaration* merely tells the *compiler* that a variable exists somewhere that has the corresponding name and type. However, this variable is not created by the declaration. If the variable is only declared, but never defined, it cannot be used. The *definition* ensures that the variable with the specified type and name is actually created somewhere in memory so that it can be worked with. With the help of *initialization, a* variable can be given a defined start value when it is defined.

Number Systems

The reasoning is very easy if we look again at the process by which a number in the base *B* number system is converted into a decimal number. In this process, the individual digits of the number are multiplied in sequence by powers of base *B* and added up. Generally speaking, the formula is

$$\left\{ z_{n-1} z_{n-2} \cdots z_2 z_1 z_0 \right\}_{B_{10}}$$
$$= z_{n-1} \cdot B_{10}^{n-1} + z_{n-2} \cdot B_{10}^{n-2} + \ldots + z_2 \cdot B_{10}^2 + z_1 \cdot B_{10}^1 + z_0 \cdot B_{10}^0$$

where the numbers to z_{n-1} represent the digits of the z_0 number. So for the number 10 in every number system the following applies

$$\{10\}_{B_{10}} = 1 \cdot B_{10}^1 + 0 \cdot B_{10}^0 = 1 \cdot B_{10}^1 = B_{10}$$

The Duotrigonal Number System

As with the hexadecimal number system, letters must be added to the individual digits to assign all 32 numbers to a single digit. The table reads:

Duotrigesimal	Decimal	Duotrigesimal	Decimal
V	31	F	15
U	30	E	14
T	29	D	13
S	28	C	12
R	27	B	11
Q	26	A	10
P	25	9	9
O	24	8	8
N	23	7	7
M	22	6	6
L	21	5	5
K	20	4	4
J	19	3	3
I	18	2	2
H	17	1	1
G	16	0	0

Print the Memory Requirement

In order to output the individual variable types with their memory requirements, a series of *cout* instructions must be used. The first output is always the text of the variable name with an equal sign, followed by the *sizeof* statement, which also takes the variable type as a parameter.

```
1   #include <iostream >
2
3   using namespace std;
4
5   int main ()
6   {
7       cout << "Memory requirements of variable :" << endl;
8       cout << "bool = " << sizeof(bool) << endl;
```

```
9      cout << "char = " << sizeof(char) << endl;
10     cout << "short = " << sizeof(short) << endl;
11     cout << "int = " << sizeof(int) << endl;
12     cout << "long = " << sizeof(long) << endl;
13     cout << "long long = " << sizeof(long long) << endl;
14     cout << "float = " << sizeof(float) << endl;
15     cout << "double = " << sizeof(double) << endl;
16     cout << "long double = " << sizeof(long double) << endl;
17
18     return 0;
19  }
```

Print the ASCII Codes

To implement the output, the line from Listing 5.2 can be adopted directly, in which the number 97 is converted to a value of type *char*. The variable can be replaced by a constant number. After that, the line only has to be copied a few times in succession and slightly modified.

```
1   #include <iostream >
2
3   using namespace std;
4
5   int main ()
6   {
7     cout << 97 << "\t= " << (char )97 << endl;
8     cout << 98 << "\t= " << (char )98 << endl;
9     cout << 99 << "\t= " << (char )99 << endl;
10    cout << 100 << "\t= " << (char )100 << endl;
11    cout << 101 << "\t= " << (char )101 << endl;
12    cout << 102 << "\t= " << (char )102 << endl;
13    cout << 103 << "\t= " << (char )103 << endl;
14    cout << 104 << "\t= " << (char )104 << endl;
15    cout << 105 << "\t= " << (char )105 << endl;
16
17    return 0;
18  }
```

Number System Conversion

The solutions for the tasks are as follows:

(a)

$$27 : 2 = 13 \quad R \quad 1$$
$$13 : 2 = 6 \quad R \quad 1$$
$$6 : 2 = 3 \quad R \quad 0$$
$$3 : 2 = 1 \quad R \quad 1$$
$$1 : 2 = 0 \quad R \quad 1$$

$\{27\}_{10} = \{00011011\}_2$

(b)

$\left\{\underset{\underset{D}{n}\ \underset{2}{n}}{11010010}\right\}_2 = \{D2\}_{16}$

(c)

$\left\{\underset{\underset{0110\,1010}{}}{6\ A}\right\}_{16} = \{01101010\}_2$

(d)

$\{127\}_8 = 1 \cdot 8^2 + 2 \cdot 8^1 + 7 \cdot 8^0$

$\{127\}_8 = \{87\}_{10}$

Binary Addition and Subtraction

The solutions always follow the same scheme. First, the numbers must be converted into binary numbers. Then follows a binary addition and the reverse conversion.

In both directions, if the sign is negative, do not forget the two's complement.

(a) Conversion of decimal numbers into the binary number system:

47	:	2	=	23	R	1	80	:	2	=	40	R	0
23	:	2	=	11	R	1	40	:	2	=	20	R	0
11	:	2	=	5	R	1	20	:	2	=	10	R	0
5	:	2	=	2	R	1	10	:	2	=	5	R	0
2	:	2	=	1	R	0	5	:	2	=	2	R	1
1	:	2	=	0	R	1	2	:	2	=	1	R	0
							1	:	2	=	0	R	1

$\{47\}_{10} = \{00101111\}_2$ $\{80\}_{10} = \{01010000\}_2$

Addition of the two binary numbers using the school method:

```
    0   0   1   0   1   1   1   1
+   0₀  1₀  0₀  1₀  0₀  0₀  0₀  0
    0   1   1   1   1   1   1   1
```

Conversion of the calculated binary number into the decimal number system:

$\{01111111\}_2 = 0 \cdot 2^7 + 1 \cdot 2^6 + 1 \cdot 2^5 + 1 \cdot 2^4 + 1 \cdot 2^3 + 1 \cdot 2^2 + 1 \cdot 2^1 + 1 \cdot 2^0$

$\{01111111\}_2 = \{127\}_{10}$

Comparison of the results in decimal numbers:

$47 + 80 = 127$

(b) Conversion of decimal numbers into the binary number system:

						73	:	2	=	36	R	1

$$73 : 2 = 36 \ R \ 1$$
$$36 : 2 = 18 \ R \ 0$$
$$4 : 2 = 2 \ R \ 0 \quad 18 : 2 = 9 \ R \ 0$$
$$2 : 2 = 1 \ R \ 0 \quad 9 : 2 = 4 \ R \ 1$$
$$1 : 2 = 0 \ R \ 1 \quad 4 : 2 = 2 \ R \ 0$$
$$2 : 2 = 1 \ R \ 0$$
$$1 : 2 = 0 \ R \ 1$$

$\{4\}_{10} = \{00000100\}_2 \qquad \{73\}_{10} = \{01001001\}_2$

Conversion of the negative number with the help of the two's complement:

$\{73\}_{10} \quad \{01001001\}_2$

invert $\quad \{10110110\}_2$

$+\{1\}_2 \quad \{10110111\}_2$

$\{-73\}_{10} \quad \{10110111\}_2$

Addition of the two binary numbers using the school method:

$$\begin{array}{ccccccccc} & 0 & 0 & 0 & 0 & 0 & 1 & 0 & 0 \\ + & 1_0 & 0_0 & 1_0 & 1_0 & 0_1 & 1_0 & 1_0 & 1 \\ \hline & 1 & 0 & 1 & 1 & 1 & 0 & 1 & 1 \end{array}$$

Since the result starts with a 1, it must be a negative number. This is converted to a positive number using the two's complement:

$\qquad \{10111011\}_2$

invert $\{01000100\}_2$

$+\{1\}_2 \quad \{01000101\}_2$

Convert the calculated binary number to the decimal number system (keeping in mind that the result of the calculation was negative):

$\{01000101\}_2 = 0 \cdot 2^7 + 1 \cdot 2^6 + 0 \cdot 2^5 + 0 \cdot 2^4 + 0 \cdot 2^3 + 1 \cdot 2^2 + 0 \cdot 2^1 + 1 \cdot 2^0$

$\{01000101\}_2 = \{69\}_{10}$

Comparison of the results in decimal numbers:

$4 - 73 = -69$

Chapter 6

Comparisons

If two expressions A and B *are* given in C++, the comparisons are done with logical comparison operators.

(a) A == B checks whether two expressions are equal. So the expression asks a question. This should not be confused with A = B, this sets two expressions equal. This is called value assignment.

(b) A <= B checks whether the expression A is less than or equal to the expression B.

(c) A ! = B checks whether an expression A is not equal to an expression B.

Instruction Blocks

A statement block consists of several statements enclosed in curly braces.

Comparisons

The expression is A = B used to assign a value. This means that the expression A on the left side is assigned the value B *on the* right side.

In contrast to this is A =   = B a logical expression. This expression is *true if and only if* the value of A *is* equal to the value of B. Otherwise, the expression is *false.*

Branches

The first instruction that can be used to create program branches in C++ is the *if* instruction. With the help of this statement, a logical expression can be evaluated. If the expression is *true,* other instructions can be executed than if the expression is *false.* Thus, an *if statement can be used to* distinguish between two different cases.

The second statement is the *switch-case* statement, which can distinguish between several cases. However, there is the restriction that the different cases must be distinguished by constant integer values. Comma numbers or variable expressions are not allowed. However, single text characters in *char* variables are permitted.

if-statement

First, an activity diagram should be developed for the application.

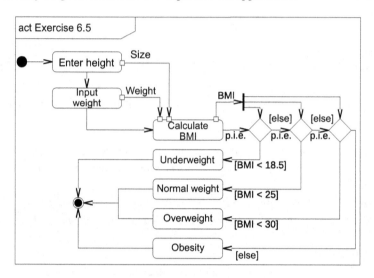

To implement the program, all that remains to be done now is to convert the diagram into code . The types and the names of the variables are already given in the task and the text of the output can be freely chosen. The formula for calculating the *body mass index* is also already included in the task.

With the case distinction at the end, care must be taken that the later cases, as can *bmi* < 25 only occur if the previous cases have already been excluded. For this reason, later *if* statements must be placed in the *else* branch of the previous *if statement.*

```
1    #include <iostream >
2
3    using namespace std;
4
5    int main ()
6    {
7        // Variable definition and initialization
8        double k = 0.0;
9        double g = 0.0;
10
```

```
11      // User input size
12      cout << "Please enter your "
13          << "body size in meters!" << endl;
14
15
16      cin >> k;
17
18      // User input weight
19      cout << "Please enter your weight"
20          << "in kilograms!" << endl;
21
22
23      cin >> g;
24
25      // Formula from the task
26      double bmi = g / (k * k);
27
28      // Case distinction
29      if (bmi < 18.5)
30        cout << "underweight" << endl;
31      else if (bmi < 25)
32        cout << "normal weight" << endl;
33      else if (bmi < 30)
34        cout << "overweight" << endl;
35      else cout << "obesity" << endl;
36
37      return 0;
38   }
```

switch-case Statement

First, an activity diagram should be developed for the application.

The task can be easily solved with the help of the activity diagram and the task definitions. Thus, the enumeration at the beginning of the program can be taken to a large extent from the task text. The variable type for the input variable is also mentioned, as well as part of the output for the user.

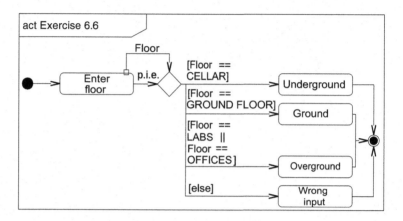

With the *switch-case* statement, two cases can be combined. This is made clear in the task by the fact that two elements of the enumeration are to produce the same output.

The hint *Wrong input!* for every other case refers to the possibility of the *switch-case* statement to process *default* cases as well.

```cpp
1   #include <iostream >
2
3   using namespace std;
4
5   // Enumeration from the task
6   enum house
7   {
8     CELLAR ,
9     GROUNDFLOOR ,
10    LABS
11    OFFICES
12  };
13
14  int main ()
15  {
16    // Variable definition and initialization
17    int e = 0;
18
19    // User input
20    cout << "Please select the floor, "
21         << "you want to go to !" << endl;
22    cout << "Cellar: " << CELLAR << endl;
23    cout << "Ground floor: " << GROUNDFLOOR << endl;
24    cout << "Labs: " << LABS << endl;
25    cout << "Offices: " << OFFICES << endl;
26    cin >> e;
```

```
27
28      // Case discrimination with switch
29      switch (e)
30      {
31      case CELLAR:
32        cout << "underground" << endl;
33        break;
34      case GROUNDFLOOR:
35        cout << "Ground level" << endl;
36        break;
37      // Since the output is identical, you can
38      // the cases are combined
39      // Fallthrough
40      case LABS:
41      case OFFICES:
42        cout << "overground" << endl;
43        break;
44      // For the case of any input
45      default:
46        cout << "Wrong input !" << endl;
47        break;
48      }
49
50      return 0;
51  }
```

Chapter 7

Loops

In *C++,* there are the following three loops, each of which belongs to the head- or foot-controlled category:

- head controlled
  - *while*-loop
  - *for*-loop
- foot.controlled
  - *do-while* loop

Use Cases

- *while-loop:* Reading a file from disk, or reaching a convergence condition in a numerical algorithm. In both cases, the termination condition is clearly defined, but the number of runs is unknown.
- *do-while loop:* A user query where not every input is allowed. Again, the termination condition is clear, but the number of passes depends on the user and is unpredictable. The loop body must also be run through at least once.
- *for-loop:* The *for* loop is a counting loop and is particularly suitable when values must be counted up or down from a certain start value with a fixed step size to a target value. Examples are outputs of function values, or working with fields. However, the latter will be discussed in the next chapter.

Loop Types

The difference between head- and foot-controlled loops lies in the time at which the termination condition is evaluated.

In a head-controlled loop, the evaluation of the termination condition takes place in the loop head. This means before the loop body is reached. For this reason, it is possible that the loop body is never run through.

In a foot-controlled loop, on the other hand, the loop body is run through first before the termination condition is checked. The loop body is therefore always run through at least once, regardless of whether the termination condition is fulfilled at the beginning or not.

Endless Loops

Each loop has a termination criterion that is evaluated either before or after the loop body. As long as the termination criterion is not fulfilled, the loop runs.

It follows that a change must occur during the run of a loop so that the termination criterion is fulfilled at some point. If this is not the case, for example, because a variable no longer changes its value or no more data is read from a file, an infinite loop is created whose termination criterion is never met.

Elevator

To achieve the desired functionality, the program only needs to be modified slightly. A *do-while* loop must be implemented around the output, the user input and the result evaluation, which runs as long as a valid input is present.

A second variable must be created in which the current floor can be stored. After the input, it must be evaluated whether the input corresponds to the current floor. Only after this evaluation may the value of the variable floor *be* changed.

```cpp
1    #include <iostream >
2
3    using namespace std;
4
5    // Enumeration from the task
6    enum house
7    {
8      CELLAR ,
9      GROUNDFLOOR ,
10     LABS ,
11     OFFICES
12   };
13
14   int main ()
15   {
16     // Variable definition and initialization
17     int e = GROUNDFLOOR ;
18     int etage = GROUNDFLOOR ;
19
20     do
21     {
22       // User input
23       cout << "Please select the floor , "
24           << "you want to go to !" << endl;
25       cout << "Cellar: " << CELLAR << endl;
26       cout << "Ground floor: " << GROUNDFLOOR << endl;
27       cout << "Labs: " << LABS << endl;
28       cout << "Bueros: " << OFFICES << endl;
29       cin >> e;
30
31       // The query checks whether the input of the
32       // current floor corresponds
33       if (e != floor)
34       {
35         // If not , the
36         // Case discrimination with switch
37         switch (e)
38         {
39           case CELLAR:
40             cout << "underground" << endl;
41             break;
```

```
42              case GROUNDFLOOR :
43                cout << "Ground level" << endl;
44                break;
45                // Since the output is identical, you can
46                // the cases are combined
47              case LABS:
48              case OFFICES:
49                cout << "overground" << endl;
50                break;
51            }
52          }
53        else
54          // Otherwise the error message
55          cout << "This is where you are right now !"
56                 << endl;
57
58        // The current floor is saved here
59        floor = e;
60      // The program runs as long as there are valid inputs.
61      } while (e >= CELLAR && e <= OFFICES );
62
63      return 0;
64    }
```

Printing of the Character Mapping Table

(a) The solution of the first part of the task can be done intuitively. Within a *for*-loop, which runs from 0 to 256, the output occurs, which is already known from Exercise 5.6.9.

```
1    #include <iostream >
2
3    using namespace std;
4
5    int main ()
6    {
7      // Variable definition and initialization
8      const int N = 256;
9
10     // for -loop for passing through the
11     // Numbers
12     for (int i = 0; i < N; i++)
13     {
14     // Output of the number and the typecast
15     // according to the familiar scheme
16     cout << i << "\t" << (char)i << endl;
17     }
```

```
18
19      return 0;
20    }
```

(b) To change the program so that the character table is output in two columns, only the *for*-loop must be replaced. In the second variant, 256 lines are no longer run through, but only half.

Two outputs must now be made for this. In the first column the normal line number can be selected. The second column is $\dfrac{N}{2}$ offset from the first by exactly one character.

```
1    // for -loop for the two-column
2    // Running through the numbers
3    for (int i = 0; i < N / 2; i++)
4    {
5      // Output of the number and the typecast
6      // according to the familiar scheme
7      cout << i << "\t" << (char)i << "\t";
8      cout << i + N / 2 << "\t"
9            << (char )(i + N / 2) << "\t";
10     cout << endl;
11   }
```

(c) With the three-column output, the principle can be continued that was already used with the two-column output. However, a problem arises, since $256/3 = 85.\overline{3}$. It follows that with a simple continuation of the principle, a number would be forgotten in the output.

The problem can be easily solved by simply outputting an additional line. However, this offset by one character must also be added to the other lines, so that the new factor results. $N/3 + 1$.

This immediately results in the next error, because two numbers are surplus in the last column. This can be prevented by using an *if* statement to prevent the number *N from being* exceeded.

Program Analysis

In order to analyze the program, it is important to understand the content of what each line of the program does. This information has been written as a comment to the lines in this version of the program.

```
1    #include <iostream >
2    // in the cmath library there are
```

```
3    // mathematical functions , e.g. sqrt
4    // to calculate the root
5    #include <cmath >
6
7    using namespace std;
8
9    int main ()
10   {
11      // Initializes a constant with the value 21.
12      // This limits the loops of the program.
13      const int N = 21;
14
15      // The names of the count variables indicate that
16      // these are x and y coordinates.
17      // The outer loop determines the line
18      for (int y = 0; y < N; y++)
19      {
20        // The inner loop determines the column
21        for (int x = 0; x < N; x++)
22        {
23          // This conversion shifts the interval
24          // of the numbers by N/2 to the top left.
25          // [0;20] becomes [ -10;10].
26          int dx = x - N / 2;
27          int dy = y - N / 2;
28
29          // sqrt calculates the square root of a number
30          // sqrt(dx * dx + dy * dy) corresponds to the theorem
31          // of Pythagoras in C++. The result is the
32          // Size of the distance of the point to the
33          // Origin. Due to the displacement
34          // which is now in the middle of the
35          // Coordinate system
36          // If the distance in the interval
37          // [N*0,1;N*0,4] lies
38          if (sqrt(dx * dx + dy * dy) < N*0.4 &&
39              sqrt(dx * dx + dy * dy) > N*0.1)
40          {
41            // a star is output
42            cout << "*";
43          }
44          else
45          {
46            // otherwise one space
47            cout << " ";
```

```
48              }
49          }
50          // Here a line is terminated
51          cout << endl;
52      }
53
54      return 0;
55  }
```

According to these considerations, the output is a circle, or an ellipse, with a hole in the middle:

Printing of the Character Mapping Table Part 2

To solve this task, you need to understand the principle after each new column shortens the number of rows that are printed.

Furthermore, an additional row must be printed if the total number, i.e. N, is not an integer divisible by the number of columns.

First, a user input must be implemented that ensures that the input is in the interval [1; 10]. A *do-while* loop is suitable for this purpose.

Now the column length must be calculated. The variable is called *factor* in this program. The form chosen in this example is quite short and exploits the fact that the result of a logical operation is always true, i.e. 1, or false, i.e. 0.

Thus, the number of rows is obtained by dividing N by the number of columns s. The modulo operation % calculates the integer remainder when dividing N by s. For 256/3, the value is 1, since the division is not smooth, and for 256 by 2, the value is 0, since 256 is divisible by 2 as an integer.

N % 2 ! = 0 Thus, the expression is true (1) if and only if N *is* not integer divisible by s and 0 otherwise.

I have to admit that this solution does not work on every system. That is why I have also included the safe variant in line 25.

The *for*-loop is supplemented by an inner *for*-loop that traverses the columns. The output is a continuation of the already known scheme.

```
1    #include <iostream >
2
3    using namespace std;
4
5    int main ()
6    {
7      // Variable definition and initialization
8      const int N = 256;
9      int s = 0;
10     int factor = 0;
11
12     do
13     {
14       // User input
15       cout << "Please specify in how many "
16            << "Columns the output should be !"
17            << endl;
18       cout << "Valid inputs: 1 - 10" << endl;
19       cin >> s;
20     } while (s < 1 || s > 10);
21
22     // Calculation of the number of lines
23     factor = N / s + (N % s != 0);
24
25     // factor = N / s;
26     // if (N % s != 0) factor ++;
27
28     // for -loop for the three-column
29     // Running through the numbers
30     for (int i = 0; i < factor; i++)
31     {
32       for (int j = 0; j < s; j++)
33       {
```

```
34            // Output of the number and the typecast
35            // according to the familiar scheme
36            if (i + j * factor < N)
37               cout << i + j * factor << "\t"
38                    << (char )(i + j * factor) << "\t";
39         }
40       cout << endl;
41     }
42
43     return 0;
44   }
```

Chapter 8

Indices

When defining an array, the number inside the square brackets indicates the number of elements.

The indices must therefore lie in the interval [0; 14].

Strings

Since the *C strings* are usually chosen so large that it is unlikely that the expected text will exceed these limits, a character is needed to indicate the end of the text within the array. In arrays of type *char,* this end is marked with the number 0. This is why we speak of null-terminated *strings.*

ASCII Table

In the ASCII table, the *American Standard Code for Information Interchange is* represented. A mapping of originally 7-bit combinations to representable characters, which was later supplemented by various extensions.

Letter Comparison

The comparison of two letters works because all characters of a text are internally represented by a number code. A comparison of two text characters therefore actually corresponds to the comparison of the two numbers. Since these have a unique mathematical relation to each other, the corresponding text characters can also be compared.

Function Values

No array is needed to solve this task!

Since the function values are not used further, a storage of the individual values is not necessary. Of course, the task could also be solved with the help of an array, but this is not necessary for the solution of the problem stated in the task.

Numbers and Characters

The character "9" is an element from the ASCII table. This means that it is a text character that does not correspond to the numerical value 9. Instead, it is assigned to the value 57 in the table.

Random Numbers

The implementation follows the task again. First, a constant $N = 100$ is defined, which is used for the definition of the array as well as for the loop passes.

Within the first loop, the array is initialized as specified in the task.

In order to be able to add up the numbers in the second loop, a variable not mentioned in the task is required in which the sum can be stored. An *int variable* could be used here, but then the decimal places would be lost during the following division. For this reason, a variable of the type *double is* suitable.

```
1    #include <iostream >
2
3    using namespace std;
4
5    int main ()
6    {
7      // Variable definition and initialization
8      const int N = 100;
9      int array[N] = {0};
10     double sum = 0.0;
11
12     // Initialization of the field
13     for (int i = 0; i < N; i++)
14     {
15       // Assign a random number
16       array[i] = edge () % 1000;
17     }
18
19     // Totaling the array elements
20     for (int i = 0; i < N; i++)
21     {
```

```
22        sum += (double)array[i];
23      }
24
25      // Division by the number of elements
26      sum /= (double)N;
27
28      // Output of the mean value
29      cout << "Mean of random numbers: "
30           << sum << endl;
31
32      return 0;
33    }
```

Largest Initial Letter

To solve the task, first the constant *N is* defined again, then a *string* array and a *string* variable for the result output.

The input of the single words is done in a *for*-loop, which shows the user how many words he still has to enter.

The value of the first element of the *words* array is assigned to the *word* variable so that it has a value entered by the user. To compensate, the search for the word with the largest initial letter can start at the second element of the array.

In the following loop, the first letter of the i-th word is compared with the first letter of the result word. If the ASCII code of the letter is larger, the result word is set to the current word, since this word obviously has the larger initial letter.

```
1    #include <iostream >
2    #include <string >
3
4    using namespace std;
5
6    int main ()
7    {
8      // Variable definition and initialization
9      const int N = 10;
10     string words[N] = {""};
11     string word = "";
12
13     for (int i = 0; i < N; i++)
14     {
15       cout << "Please enter the "
16            << i + 1 << "th word of "
17            << N << " on:"
18            << endl;
19       cin >> words[i];
```

```
20    }
21
22    // so that the variable with a
23    // valid value is initialized
24    word = words [0];
25
26    for (int i = 1; i < N; i++)
27    {
28      // if the first letter of the i-th is
29      // word is greater than that of the
30      // Result word
31      if (words[i][0] > word [0])
32        // the word should be exchanged
33        word = words[i];
34    }
35
36    // Output
37    cout << "word with the first letter ,"
38        << "furthest back in the alphabet"
39        << "is: "
40        << word << endl;
41
42    return 0;
43  }
```

Program Analysis

When analyzing the individual program sections, you can arrive at the following results:
   The loops are best understood when the analysis is done from the inside out.

```
1   #include <iostream >
2   // the library time.h contains functions ,
3   // with which time information is requested and
4   // can be saved
5   #include <time.h>
6
7   using namespace std;
8
9   int main ()
10  {
11    // srand initializes the random number generator ,
```

```
12      // which normally always creates the same random -
13      // numbers.
14      // time (0) returns the current time in
15      // seconds since 1.1.1970
16      srand(time (0));
17
18      // Initialization of a double array
19      // with 1000 elements
20      const int N = 1000;
21      double values[N] = { 0.0 };
22
23      // Here are the values of the array
24      // calculated
25      for (int i = 0; i < N; i++)
26      {
27        // rand () % 1000 returns a whole number
28        // in the interval [0;999] , this is written as double
29        // and then divided by 100
30        // the result is a number in the interval
31        // [0.00;9.99]
32        values[i] = (( double )(margin () % 1000)) / 100.0;
33      }
34
35      // (4) This loop lets N times the current largest
36      // element move to the end
37      for (int i = 0; i < N; i++)
38      {
39        // (3) This loop ensures that the
40        // largest element of the array moves to the end
41        //
42        for (int j = 0; j < N - 1; j++)
43        {
44          // (1) Two elements are compared here ,
45          // which are adjacent to each other
46          // it is checked if the left element is larger
47          // than the right
48          if (values[j] > values[j + 1])
49          {
50            // (2) the two are interchanged
51            double h = values[j];
52            values[j] = values[j + 1];
53            values[j + 1] = h;
54          }
55        }
56      }
```

```
57
58     // the values are output here
59     for (int i = 0; i < N; i++)
60     {
61        cout << values[i] << endl;
62     }
63
64     return 0;
65   }
```

If all the findings are summarized, it can be concluded that the program creates a series of random numbers and then sorts them in ascending order.

The procedure is known in computer science under the name *Bubblesort*, because the largest number in each case, like an air bubble in the water rises to the top. However, the algorithm is rarely used in practice, because there are faster methods for sorting numbers.

Word Lengths

To solve this task, the knowledge from several previous chapters must be combined. In addition, a little research of your own is required.

Numbers in the interval [3; 10] must be rolled out to determine the word lengths. This requires the constant number 3, to which a random number in the interval [0; 7] must be added. To calculate random lowercase letters, this principle must be applied again. For this, however, it must first be recognized that a random lowercase letter corresponds to a random number in the interval [97; 122].

After initialization of the array, all words have length 0, so direct access to the first or second letter is not possible. Instead, letters must be appended to the end of the word. This is done either with the operator +, or, as in this example, with the operator +=, which combines addition and value assignment.

```
1    #include <iostream >
2    #include <string >
3    #include <time.h>
4    #include <cmath >
5
6    using namespace std;
7
8    int main ()
9    {
10       // Initialization of the
11       // Random number generator
```

```
12      srand(time (0));
13
14      // Variable definition and initialization
15      const int N = 1000;
16      string words[N] = {""};
17      double x = 0.0;
18      double s = 0.0;
19
20      for (int i = 0; i < N; i++)
21      {
22        // Here the word lengths are diced
23        // The minimum length is 3, so
24        // this value is firmly added up , then
25        // a random number must be added, which is
26        // Maximum 7 is , so rand () % 8
27        int l = 3 + edge () % 8;
28        for (int j = 0; j < l; j++)
29        {
30          // Lower case letters must be used here
31          // to be diced
32          // They lie in the interval [97;122]
33          // Since the words do not yet have a length
34          // the letters must be separated by +=
35          // to be appended
36          words[i] += (char )(97 + margin () % 26);
37        }
38        // Here the lengths are added up ,
39        // in order to find the mean value in the first loop.
40        // to be calculated
41        x += l;
42      }
43      // Completion of the mean value calculation by
44      // Division with the number of elements
45      x /= N;
46
47      for (int i = 0; i < N; i++)
48      {
49        // Part of the calculation of
50        // Standard deviation
51        // pow(a,b) calculates the value a to the power of b
52        s += pow(words[i]. length () - x, 2);
53      }
54      // Completion of the calculation of
55      // Standard deviation
56      s /= N - 1;
```

```
57      s = sqrt(s);
58
59      // Output of words
60      cout << "words :" << endl << endl;
61      for (int i = 0; i < N; i++)
62      {
63        cout << words[i] << endl;
64      }
65
66      // Output of results
67      cout << "Average word length: "
68           << x << endl;
69      cout << "Standard deviation of word lengths: "
70           << s << endl;
71
72      return 0;
73    }
```

Chapter 9

Function Prototype

To create a functional prototype, three pieces of information are needed:

- The return type of the function
- The name of the function
- The function parameters (number and type)

Return Value

Although the function *f* has no return type, *void* must still be specified.

No names were given for the function parameters. However, these are important information for a function header, so that the parameters can also be used within the function. In fact, it would also be permissible to omit them in the function header.

So one possible solution is:

```
void f(int a, double b, char c)
{}
```

Call by Reference

The term *Call by Reference* describes the way variables are passed to a function. Normally, new variables are created for each function parameter and the passed values are copied into these new variables. If the copy is changed, the original remains untouched. This is called *Call by Value*.

In a *call by reference, the* name of the function parameter is preceded by an ampersand. Now no new variable is created, instead the new name refers to the already existing variable. If the value of the variable is now changed within the function, this corresponds to accessing the original value, which consequently changes as well.

Variadic Functions

The term "variadic function" describes a function whose parameters are not uniquely defined either in number or in type. Instead, the function can accept a variable number of parameters. However, it is neither possible to determine the number nor the type of the parameters by itself. For this reason, additional parameters or restrictions are necessary to evaluate the parameters.

In order to implement variadic functions, the file *stdarg.h* must be included with the *preprocessor directive #include.*

Recursion

In a recursive solution, a function f *is* implemented consisting of a recursive and a non-recursive path. The recursive path calls the function f again with changed parameters, while the non-recursive path interrupts the series of self-calls at certain conditions that depend on the task. The recursive solution thus exploits the self-similarity of certain actions.

Iteration describes a process that can be solved with the help of a loop. Here, too, a principle is applied to a data set again and again.

In principle, suitable problems can be solved recursively as well as iteratively. However, the effectiveness of both approaches is strongly dependent on the underlying problem.

static

A variable that is provided with the attribute *static* within a function is, in contrast to normal variables, only created and initialized once within the function. Thus, this variable is able to store information over several function calls. For example, it can be used to cause a function to count how many times it has been called.

Function Overloads

In *C++* it is possible to define several functions with the same name if they have different parameter configurations. However, a distinction only in the return type is not permitted!

If it can happen that information is available in different forms, such as a text that can be available as a *string* or as an array of type *char*, then two different functions can be offered that can react optimally to the available data types without having to perform a type conversion.

Input and Output Functions

In solving this task, the new *getline* statement is used, with which entire lines including spaces can be read in from the console and stored in a *string*.

The *input*-function must realize an input, therefore it does not need any parameters, but returns the read value.

The *output* function does not have to return anything, but it needs the text to be printed as a parameter.

Since a dashed line is to be printed before and after the text, the *help* function was implemented. This was not explicitly mentioned in the task, however, this recurring task fulfills exactly the requirements for the creation of a function.

In the main function, you only have to go through all the characters of the text to replace each space with an asterisk.

```
1    #include <iostream >
2    #include <string >
3
4    using namespace std;
5
6    string input ()
7    {
8      // Variable definition and initialization
9      string text = "";
10
11     // Input according to the task
12     cout << "Please enter text".
13         << endl;
14     getline(cin , text );
15
16     return text;
17   }
18
19   // Auxiliary function
20   void help(int n)
21   {
22     // Output of n hyphens
```

```
23     for (int i = 0; i < n; i++)
24     {
25       cout << "-";
26     }
27     cout << endl;
28   }
29
30   void output(string text)
31   {
32     // Calling the auxiliary function
33     // with the text length
34     help(text.length ());
35     // Text edition
36     cout << text << endl;
37     // The auxiliary function again
38     help(text.length ());
39   }
40
41   int main ()
42   {
43     // Variable definition and initialization
44     // with the input function
45     string text = input ();
46
47     // Replacing spaces with asterisks
48     for (int i = 0; i < text.length (); i++)
49     {
50       if (text[i] == ' ')
51         text[i] = '*';
52     }
53
54     // Output
55     output(text);
56
57     return 0;
58   }
```

Recursion

The implementation of the program is simple if the instructions from the task are followed. The function *recursion* is to return nothing, hence the return type *void*. *The* function parameter *c is* to be of type *int* and is named in the task definition.

Within the function, a self-call is to occur if the value of *c* is less than 100.

In the main function, only the recursive function should be called.

In task part (a), the output is to follow the recursive branch. As a result, the numbers are counted down from 100 to 0 on the screen.

For task part (b), the value should be printed before the recursive branch. Suddenly the numbers are incremented from 0 to 100.

This happens for the following reason: In task part (a), the first call to the function runs into the recursive branch and calls itself with the value 1. The same happens with the following functions until c takes the value 100.

Only then does the last function call stop running into the recursive branch and reach the output and exit. The penultimate function call has then terminated the recursive branch and also reaches the output, and so on. The numbers are counted down from 100.

In the task part (b) it runs similarly, only that here the output takes place before the recursive branch. For this reason, the numbers are incremented.

```
1    #include <iostream >
2
3    using namespace std;
4
5    void recursion(int c)
6    {
7      // Output of the parameter
8      // Task part (b)
9      // cout << c << endl;
10
11     // recursive branch , if
12     // c < 100
13     if (c < 100)
14       recursion(c + 1);
15
16     // Output of the parameter
17     // Task part (a)
18     cout << c << endl;
19   }
20
21   int main ()
22   {
23     // Calling the recursive function
24     recursion (0);
25
26     return 0;
27   }
```

Program Analysis

When analyzing this program, it makes sense to start at the beginning of the program flow, i.e. at the main function. If it is clear what happens there, the examination of the further functions can take place. But even with these functions, it makes sense to stick to the order given by the program flow.

```
1    #include <iostream >
2
3    using namespace std;
4
5    int func(int val[], int s, int e)
6    {
7      // The function terminates itself ,
8      // if the first and last index are identical
9      // In this case the
10     // Indexes describe exactly one element.
11     // Its value is returned.
12     if ((e - s) == 0) return val[s];
13
14     // Here the index is calculated that
15     // lies exactly between the first and the
16     // last index.
17
18     // Because of the integer division
19     // always the smallest value is assumed
20     int h = (e + s) / 2;
21
22     // The function calls itself twice
23     // once with the lower half of the array
24     // and once with the upper half
25     int e1 = func(val , s, h);
26     int e2 = func(val, h + 1, e);
27
28     // Finally, the results are added together.
29     return e1 + e2;
30   }
31
32   int main ()
33   {
34     // Here, an array with N = 100
35     // Values is created
36     const int N = 100;
37     int values[N];
38
```

```
39      // Within the loop, the
40      // Array with the numbers 1 to N
41      // is initialized
42      for (int i = 0; i < N; i++)
43      {
44        values[i] = i + 1;
45      }
46
47      // Finally, the result of the
48      // unknown function is output
49      // The parameters are the array , and the
50      // first and last valid index
51      cout << "Result: " << func(values , 0, N - 1)
52           << endl;
53
54      return 0;
55    }
```

Since the program splits the array up to one-element arrays and then adds the values in sequence, the result must therefore be a sum. Since each value is added only once (even if the value occurs several times in sums in the form of a partial result, of course), the result must therefore be the sum of the numbers from 1 to 100.

Result: The program calculates the value 5050.

Output of Parameters of a Variadic Function
The solution of this task is not extensive, but understanding the variadic functions is not easy. Here it is especially important to follow the sequence of steps to solve the task.

First, the *stdarg.h* file must be included so that the required functions are available.

The parameter list must be defined as a variable of type *va_list* and initialized by the function *va_start*. Here especially the second parameter is important, because here the name of the *string* must be given, after which the free parameters begin.

During the evaluation the function *va_arg* must be used to extract the parameters from the list.

Finally, the list must be released again by *va_end*.

```
1    #include <iostream >
2    #include <string >
3    #include <stdarg.h>
4
5    using namespace std;
```

```
6
7    void myPrint(string t, ...)
8    {
9      // Definition of the variable list
10     va_list params;
11     // Initialization of the variable list
12     // after the parameter t
13     va_start(params , t);
14
15     // Runs through the passed text
16     for (int i = 0; i < t.length (); i++)
17     {
18       // the sign is not a star
19       if (t[i] != '*')
20         // this is how it is output
21         cout << t[i] << endl;
22       else
23       {
24         // otherwise the next
25         // Parameters output in the list
26         cout << va_arg(params , int)
27               << endl;
28       }
29     }
30
31     // Finally, the list must be
32     // released
33     va_end(params );
34   }
35
36   int main () {
37     // Calling the variadic function
38     myPrint ("-+-*-+-*-+-*", 1, 2, 3);
39
40     return 0;
41   }
```

If the program is executed, the output is:

```
-
+
-
1
-
+
```

−
2
−
+
−
3

Chapter 10

Visibility Levels

Three different visibility levels can be applied in classes:

- **public:** All variables and functions defined with this visibility level are accessible from outside as well as from inside the class.
- **protected:** The visibility level *protected* protects variables and functions from access from outside the class.
- **private:** In addition to protecting against access from outside the class, the visibility level *private* prevents the corresponding members from being inherited.

Operators

Operators can be used to perform standard arithmetic, comparison and logic operations in classes. The operators are special functions that allow, for example, to use the usual formulations in mathematics, such as $A + B$.

However, operators can also be used to add functions to existing classes. For example, an operator can be used to extend the *cout* class so that objects of your own classes can be printed to the console.

Include Guards

An *include guard* is set up by preprocessor directives. First, the *#ifndef* directive is used to check whether a certain term has already been defined. If not, the term is defined by #define to prevent the corresponding area from being reached a second time.

Since the *header* files of classes can be included in several other files, this would have the consequence that the corresponding classes cannot be compiled without *Include-Guard*. With each inclusion it would be tried again to define the class.

Abstract Classes

In *C++,* an abstract class designates a class in which at least one function has been defined as virtual and provided with the addition =0. This has the consequence that the corresponding function is set as part of the interface of the class, but is not defined in the current class. This must be done in an inheriting class.

No object can be created from the class itself in this way, the class is "abstract".

Member Variables

The *member variables* are specified when the class itself is defined. They are available in every function of the class and store information even across function boundaries. Their lifetime is directly bound to the object of the class.

All other variables can only be created as parameters or within functions. The lifetime of these variables is bound to the respective function.

Constructors

Several constructors can exist within a class. Exactly one constructor is always called when a new object of the class is created. This distinguishes constructors from all other functions. A constructor is guaranteed to be called when an object is created, but never again after that for this object.

Classes and Structures

In the *C++* language, apart from the name of the construct, there is only one other difference between classes and structures. The default visibility level of the class is *private,* that of the structure is *public.*

However, there is a perceived difference for many programmers. Structures are data containers that have few, if any, functions and usually leave the data at the *public* visibility level.

Classes are more complex entities that protect their data and can perform many different tasks with their data.

Strictly speaking, however, this distinction in functionality does not exist.

Polymorphism

The term polymorphism means multiformity and is very closely related to the term inheritance. It describes the property of functions in different classes that inherit from each other to appear with the same name, but always behave differently.

This is related to the *virtual* keyword, which allows existing functions to be reimplemented in inheriting classes. Since functions that have inherited from another class can also be passed by reference to their base class, calling a function can have a different effect depending on which class was passed.

Polymorphism

Normally, variables and functions are bound to an object of a class. This means that values are stored for an object and functions are called on a concrete object whose values are then worked with.

If, on the other hand, a variable or a function is defined as *static,* these elements are bound to the class itself and can be used without a concrete object of the class having to exist. Because of this property, static variables have the same value across all objects of a class. This makes it possible, for example, to count the number of objects in a class. To do this, you only need to create a static variable that is incremented in each constructor and decremented in each destructor. The value of the variable then always corresponds exactly to the number of currently existing objects.

The Class *point2D*

The solution of this task can be taken over to a large extent from the problem definition, if the terms, like constructor or operator were understood.

The formulas for the mathematical functions are also given in the task, so that these only have to be translated into *C++ code.*

The example has great similarity to the *Vector2D* class developed in the book.

In the solution, every reference passed and every function was defined as constant, if this was possible. But this is only to give an impression how the keyword *const* can be used in programs.

Solutions that do not apply this consistently should still be considered correct (Listings 1 and 2).

Listing 1 *point2D.h*

```
1    // Include -Guard
2    #include <iostream >
3
4    using namespace std;
5
6    class point2D
7    {
8    public:
9      // Constructor without parameters
10     point2D ();
11     // Constructor for initialization

12     point2D(double x, double y);
13     // Copy Constructor
```

```
14      point2D(const point2D &p);
15
16      // Addition
17      point2D operator +( const point2D &p) const;
18      // Scalar product
19      double operator *( const point2D &p) const;
20      // Output via cout
21      friend ostream& operator <<(ostream &out ,
22                                      const point2D &p);
23   protected:
24      double m_x;
25      double m_y;
26   };
27   // Output via cout
28   ostream& operator <<(ostream &out , const point2D &p);
```

Listing 2 *point2D.cpp*

```
1    #include "point2D.h"
2
3    // Initialization to the origin
4    point2D :: point2D ()
5      : m_x (0)
6      , m_y (0)
7    {}
8    // Individual initialization
9    point2D :: point2D(double x, double y)
10     : m_x(x)
11     , m_y(y)
12   {}
13   // Copy of an existing point2D
14   point2D :: point2D(const point2D &p)
15     : m_x(p.m_x)
16     , m_y(p.m_y)
17   {}
18   // The coordinates of the result point
19   // result from the sum of the coordinates
20   // of the points
21   point2D point2D :: operator +( const point2D &p) const
22   {
23     point2D result;
24
25     result.m_x = m_x + p.m_x;
26     result.m_y = m_y + p.m_y;
```

```
27
28    return result;
29  }
30  // Scalar product calculated according to
31  // specified formula
32  double point2D :: operator *( const point2D &p) const
33  {
34    return m_x * p.m_x + m_y * p.m_y;
35  }
36  // Output function
37  ostream& operator <<(ostream &out , const point2D &p)
38  {
39    out << "point2D (" << p.m_x << ", "
40      << p.m_y << ")";
41
42    return out;
43  }
```

The *Circle* Class

In this task it is important to understand how own classes can be used in other classes. This allows complex problems to be distributed among different classes and thus simplified.

The class *point2D* already does all the tasks that need to be implemented for the point. The output is already there and also for the initialization different possibilities are already available. In fact, even the initialization could be omitted completely if the coordinates (0, 0) are to be selected.

The calculation of the area and the circumference can be taken directly from the task. These functions can also be provided with the keyword *const*, since the properties of the circle are not changed during the area calculation.

When outputting the circle on the console, the output function of the point can be applied directly. This simplifies the output function of the circle (Listings 3 and 4).

Listing 3 *circle.h*

```
1   #include "point2D.h"
2
3   class circle
4   {
5   public:
6     // Constructor without parameters
7     circle ();
8     // Initialization with single values
9     circle(double x, double y, double r);
10    // Initialization with the aid of a
11    // point2D and a value
```

```
12    circle(const point2D &p, double r);
13    // Calculation of the area
14    double area () const;
15    // Calculation of the scope
16    double perimeter () const;
17    // Output on the console
18    friend ostream& operator <<( ostream &out , const circle &c);
19  protected:
20    point2D m_center;
21    double m_r;
22  };
23
24  ostream& operator <<( ostream &out , const circle &c);
```

Listing 4 *circle.cpp*

```
1   #include "circle.h"
2
3   // Definition of a constant for Pi
4   const double PI = 3.14159265358979323846426433832795;
5   // Initialization of a unit circle in the origin
6   // In each constructor, any of the constructors defined in point2D
7   // defined constructors can be used ,
8   // to initialize the variable m_center
9   circle :: circle ()
10    : m_center (0, 0)
11    , m_r (1)
12  {}
13  // individual initialization
14  circle :: circle(double x, double y, double r)
15    : m_center(x, y)
16    , m_r(r)
17  {}
18  // individual initialization
19  circle :: circle(const point2D &p, double r)
20    : m_center(p)
21    , m_r(r)
22  {}
23  // Calculation of the area according to the task
24  double circle ::area () const
25  {
26    return PI * m_r * m_r;
27  }
28  // Calculation of the scope according to the task
```

```
29    double circle :: perimeter () const
30    {
31      return 2 * PI * m_r;
32    }
33    // During output, the output function
34    // implemented in point2D can be used.
35    // was
36    ostream& operator <<(ostream &out , const circle &c)
37    {
38      out << "circle (" << c.m_center << ", "
39        << c.m_r << ")";
40
41      return out;
42    }
```

Program Analysis

The analysis of this program is a bit more difficult than before, because there is no main program where the analysis could start. Instead there is a *header* file and a *.cpp* file (Listings 5 and 6).

Listing 5 *Riddle.h*

```
1    // Include -Guard
2    #include <iostream >
3    #include <string >
4
5    using namespace std;
6
7    class Riddle
8    {
9    public:
10     // Constructor
11     Riddle(string data );
12
13     // the output may access the protected
```

```
14       // elements of the class
15       friend ostream& operator <<(ostream &out , Riddle r);
16    protected:
17       // The class stores only one text
18       string m_data;
19    };
20    // Output of the text with an ostream
21    ostream& operator <<(ostream &out , Riddle r);
```

Listing 6 *Riddle.cpp*

```
1     #include "Riddle.h"
2
3     Riddle :: Riddle(string data)
4     {
5       char k;
6
7       // The loop passes through all letters
8       // of the transferred text
9       for (int i = 0; i < data.length (); i++)
10      {
11        // this is the current letter
12        k = data[i];
13
14        // If it is a
15        // Lower case letter
16        if (k >= 97 && k <= 122)
17          // k - 94 shifts the interval
18          // [97;122] to [3;28] the modulo
19          // truncates the trailing characters
20          // and pushes it to the beginning.
21          // Now all characters are in the interval
22          // [0;25]. If 65 is added, they become
23          // Capital letters
24          k = 65 + (k - 94) % 26;
25        else
26          // they are already capital letters
27          if (k >= 65 && k <= 90)
28            // we also shift
29            // but the letters remain
30            // capital letters
31            k = 65 + (k - 62) % 26;
32
```

```
33        // other characters are simply copied
34        m_data += k;
35      }
36   }
37
38   ostream& operator <<(ostream &out , Riddle r)
39   {
40     char k;
41
42     for (int i = 0; i < r.m_data.length (); i++)
43     {
44       k = r.m_data[i];
45
46       // When the program writes its output
47       // there are only capital letters
48       if (k >= 65 && k <= 90)
49         // these are taken from the interval [65;90]
50         // shifted into the interval [23;48].
51         // once again, the modulo
52         // joins the end and the beginning
53         k = 65 + (k - 42) % 26;
54
55       out << k;
56     }
57
58     return out;
59   }
```

However, the advantage of this program is that there is only one constructor and one function, so the order of calls is fixed. Again, the analysis should start from the beginning. So in this case with the constructor.

The program encrypts text by turning each letter into an uppercase letter and shifting it three places to the right in the alphabet. This very old and insecure encryption is called *Caesar cipher.*

Only the encrypted texts are stored within the data structure. Only for decryption the text is converted back into plain text. To do this, all letters are shifted 23 places to the right and placed back at the beginning of the alphabet by the modulo.

This corresponds to a shift of three places in the alphabet to the left.

Chapter 11

Memory Areas

The memory of a program is divided into the following four areas:

- **Program code:** This area contains the executable code of the program.
- **global variables:** The global variables get their own memory area, because they differ in behavior from other variables.
- *Stack:* All information needed for the execution of a function is stored on the *stack,* i.e. local variables, function parameters and the return address.
- *Heap:* The *heap is* needed for dynamic memory requests. The program can request new memory areas on the *heap at* any time using statements such as *new* or *malloc,* but must manage these itself. If requested memory areas are forgotten and not released, the memory can fill up.

Dereferencing

Dereferencing refers to indirect access to a value via a pointer. Normally, a value is stored in a variable that is located at a specific address. With a pointer, however, another address is stored instead of the value, at which the value is then located.

Dereferencing first jumps to the stored address, then gets the value.

Multidimensional Arrays

The following three options were presented:

- **Pointers to arrays:** With $C++$, it is possible to create pointers to fixed-size arrays. In this way, a one-dimensional array can be created from pointers whose elements in turn point to fixed-size arrays. With this solution, however, part of the array remains on the *stack.*
- **Pointer to pointer:** In this variant, a double pointer must be created, i.e. a pointer that points to pointers. Then an array of pointers is created on the *heap,* which can then be initialized individually with their own arrays in a loop. This variant is very flexible, but also means a high administration effort and a deeper understanding of pointers.
- **virtual dimensions:** It is possible to create a one-dimensional array on the *heap* and to create the other dimensions virtually by mathematical formulas itself. For this purpose, an array of size $Y \times X$ is created and then divided into Y pieces of length X *by the* formula *index* $= x + y \cdot X$.

Function Pointer

A function pointer can be created using the *typedef* instruction. However, this is only required if a new variable type is to be created.

The information needed to create a function is always the same, therefore it is also needed for a function pointer:

- **Return type**
- **Name** – For a function pointer to be created, the name must be enclosed in parentheses and begin with an asterisk.
- **Function parameters**

Stack and *Heap*

Information required for the execution of functions is stored on the stack memory. This information includes the local variables, the function parameters and the return address. Since these tasks are of a very basic nature, a definable but fixed contingent of memory is available for each program. This memory is organized according to the *LIFO* principle, which means that the information that was stored last is the first to leave the memory. Accesses to the *stack* can be made very quickly because of this ordered structure.

The memory on the *heap is* only available when it is requested by the program at run-time. The program can decide when and how much memory it requests. In addition, it can release requested memory at any time. Memory can fragment if memory is requested and released frequently and the size of the requested areas varies. Also, because of its dynamic nature, accesses to the *heap* are slower than to the *stack*. On the other hand, the size of the requested memory is in principle not limited (the resource itself is of course limited).

Memory Consumption

If an image consists of 1024 × 768 pixels, then the number of all pixels is given by 1024 · 768 = 786.432 pixels. If a pixel consists of 16 bits, i.e. 2 bytes, then 786432 · 2 = 1572.864 bytes are needed to store the image. This corresponds to 1536 kilobytes, or 1.5 megabytes.

In computer science, a kilobyte does not consist of 1000 bytes, but of 1024 bytes.

Pointer Arithmetic

With pointer arithmetic, it should be noted that the size of the data type of the pointer is always assumed as the unit. If the value 1 is added to an integer, the result is larger by the value 1, as expected.

If the value 1 is added to a pointer of type *int*, the address contained in it is not increased by the value 1, but by the value 1 · 4 bytes, the size of the stored data type *int*.

Memory Reservation

Some implementations of the *new* instruction use the *malloc* instruction internally *to* reserve memory. In this case, the memory can also be released again by *free*. However, this is by no means guaranteed. To make matters worse, the *new* statement can be overwritten as an operator at any time. This can happen for a variety of reasons, for example, to implement your own memory management (this is not uncommon in time-critical applications). The functionality of the *new* statement is therefore not clearly defined.

For this reason, it is unsafe to mix the different instruction groups and should always be avoided.

Random Numbers

To solve the task, the random number generator must first be initialized. The user query for the array size must be made early in the program, since the array cannot be created until the result of the input is available.

While the array is initialized with random numbers, the values can be summed up in parallel in the variable *x to* prepare the calculation of the expected value.

In a second loop, the standard deviation can then be calculated.

At the end, the results must be printed and the memory released again. It is important to use the statement *delete[],* since the memory for an array is to be released.

```cpp
1    #include <iostream>
2    #include <cmath >
3    #include <time.h>
4
5    using namespace std;
6
7    int main ()
8    {
9      // Initialization of the random number -
10     // generator
11     srand(time (0));
12
13     // Variable definition and initialization
14     int N = 0;
15     double x = 0.0;
16     double s = 0.0;
17     int *values = 0;
18
19     // User query with interval limits
20     do
21     {
22       cout << "Please enter an integer "
23            << "number between 1 and 1000 :";
24       cin >> N;
25     } while (N < 1 || N > 1000);
26
```

```
27      // Create the array. only now
28      // the size of the array is known.
29      values = new int[N];
30
31      // Loop for array initialization
32      for (int i = 0; i < N; i++)
33      {
34        // Generation of random numbers
35        // in the interval [1;6]
36        values[i] = 1 + edge () % 6;
37        // Summing the values for the
38        // mean value
39        x += values[i];
40      }
41      // Division by the number of elements
42      x /= N;
43
44      // Calculation of the standard deviation
45      for (int i = 0; i < N; i++)
46      {
47        // Summing the squared errors
48        s += pow(values[i] - x, 2);
49      }
50      // Division and extraction of the root
51      s /= (N - 1);
52      s = sqrt(s);
53      // Output of results
54      cout << "Mean value: " << x
55          << endl;
56      cout << "Standard deviation: "
57          << s << endl;
58      // Releasing the memory and deleting it
59      // the address
60      delete [] values;
61      values = 0;
62
63      return 0;
64    }
```

Random Numbers the Second

The first steps of this program behave very similarly as in the solution before, so that they should not be explained here again.

By choosing a one-dimensional array that is virtually split into several dimensions by applying a formula, the programmer gains additional freedom.

Tasks that are performed the same way for all elements do not have to be solved by multiple loops, but can be solved by a single loop. A good example is the value initialization. All values are to be selected randomly in the interval [1; 6]. A division into rows and columns is not necessary for this, therefore a single loop is sufficient.

The calculation of the expected value also behaves similarly. The elements of a row would have to be summed up to get the value of the row. After that, the results would have to be summed up to calculate the arithmetic mean. The result is identical to the sum of all the elements of the array. However, when dividing, care must be taken to divide only by the number of rows, as the results should be summed row by row.

When calculating the standard deviation, two loops must actually be used. The sum of the elements of each line is needed for the calculation of the squared error to the expected value.

```cpp
1    #include <iostream >
2    #include <time.h>
3    #include <cmath >
4
5    using namespace std;
6
7    int main ()
8    {
9      // Initialization of the
10     // Random number generator
11     srand(time (0));
12
13     // Variable definition and initialization
14     int X = 0;
15     int Y = 1000;
16     double x = 0.0;
17     double s = 0.0;
18     int sum = 0;
19     int *values = 0;
20
21     // User query with interval limits
22     do
23     {
24       cout << "Please enter an "
25            << "integer in the interval "
26            << "1 to 10:" << endl;
27       cin >> X;
28     } while (X < 1 || X > 10);
29
30     // Create the array. only now
31     // the size of the array is known.
```

```
32      values = new int[X*Y];
33
34      // To initialize all values it is sufficient to
35      // do a single loop over all values
36      // The mean value can also be
37      // calculated, since for the arithmetical
38      // mean all values are summed
39      //
40      for (int i = 0; i < X*Y; i++)
41      {
42        // Generation of random numbers
43        // in the interval [1;6]
44        values[i] = 1 + edge () % 6;
45        // Summing the values for the
46        // mean value
47        x += values[i];
48      }
49      // Important!
50      // Division by the number of lines
51      x /= Y;
52
53      // The calculation of the standard deviation
54      // must be done in two loops
55      for (int i = 0; i < Y; i++)
56      {
57        sum = 0;
58        for (int j = 0; j < X; j++)
59        {
60          // first, the values of a line
61          // must be summed up
62          sum += values[j + i * X];
63        }
64        // Then the squared error
65        // is calculated
66        s += pow(sum - x, 2);
67      }
68      // then the division takes place
69      // and the root is calculated
70      s /= (Y - 1);
71      s = sqrt(s);
72
73      // Output of results
74      cout << "Mean value: " << x
75            << endl;
76      cout << "Standard deviation: "
```

```
77          << s << endl;
78    // Releasing the memory and deleting
79    // the address
80    delete [] values;
81    values = 0;
82
83    return 0;
84  }
```

Program Analysis

The explanation for the individual program steps was written as a comment to the lines.

```
1   #include <iostream >
2
3   using namespace std;
4
5   int main ()
6   {
7     // Variable definition and initialization
8     // three variables are normal stack variables
9     // the third lies on the heap
10    int x = 0;
11    int y = 0;
12    int k = 0;
13    double *z = new double (3.0);
14
15    // Since z is a pointer, z contains the address
16    // which was reserved on the heap
17    // This address is converted to an integer
18    // by the typecast and saved in y
19
20    y = (int)z;
21
22    // The dereferencing *z returns the content
23    // of the double , which was created on the heap
24    // this is represented by a typecast
25    // as integer and stored in k
26    // k is now 3;
27    k = (int)*z;
28
```

```
29      // y is on the stack , therefore &y returns
30      // the stack address at which y is located. This
31      // is converted to an integer by a typecast
32      // and stored in x.
33      x = (int)&y;
34
35      // The value stored in y is equal to
36      // the address of y, but is an integer.
37      // The typecast causes the address value to be
38      // interpreted as double.
39      // This works , because the address is valid. Then
40      // the pointer is dereferenced by the preceding asterisk
41      // The result is the numerical value
42      // in y, i.e. 3. This value is multiplied by 2
43      // and stored.
44      // The heap now contains the value 6
45      *(( double *)y) *= 2;
46
47      // The dereferencing of z now returns the value
48      // 6, since it was changed in the step before this one
49      // in k there is the value 3, so the value 2 is now stored
50      // at the address to which z points
51
52      *z /= k;
53
54      // Now the whole chain. In x is the address of
55      // y, this is again interpreted as an address
56      // and dereferenced. The result
57      // is the value in y. This is the address of z
58      // this is used as a pointer of type double
59      // and also dereferenced.
60      // The result is the number 2.0
61      // This is used as an integer and
62      // issued.
63      cout << (int )*(( double *)(*( int*)x)) << endl;
64
65      // Finally, the memory is released and
66      // the address is deleted.
67      delete z;
68      z = 0;
69
70      return 0;
71  }
```

The output is therefore 2!

References

Apple Distribution International. (2017). https://itunes.apple.com/de/app/xcode/id497799835?mt=12

Bloom, B., Engelhart, M., Furst, E., Hill, W., & Krathwohl, D. (1956). *Taxonomy of educational objectives. The classification of educational goals, handbook I: Cognitive domain.* David McKay Company.

Bousseljot, R., Kreiseler, D., & Schnabel, A. (1995). Nutzung der ekgsignaldatenbank cardiodat der ptb über das internet. *Biomedizinische Technik, 40*(1), 317.

Code::Blocks. (2017). http://www.codeblocks.org/

Cooley, J. W., & Tukey, J. W. (1965). An algorithm for the machine calculation of complex fourier series. *Mathmatics of Computation, 19*(90), 297–301.

Eclipse Foundation. (2017). https://www.eclipse.org/cdt/

Gertsch, M. (2008). *Das EKG* (2nd ed.). Springer.

Goldberger, A., Amaral, L., Glass, L., Hausdorff, J., Ivanov, P., Mark, R., Mietus, J., Moody, G., Peng, C.-K., & Stanley, H. (2000). Physiobank, physiotoolkit, and physionet: Components of a new research resource for complex physiologic signals. *Circulation, 101*(23), e215–e220. http://circ.ahajournals.org/content/101/23/e215.full

Goovaerts, H. G., Ros, H. H., vanden Akker, T., & Schneider, H. (1976). A digital QRS detector based on the principle of contour limiting. *IEEE Transactions on Biomedical Engineering, BME-23,* 154.

Lichtenberg, G., & Reis, O. (2016). Kompetenzgraphen zur Darstellung von Prüfungsergebnissen. In B. Berendt, A. Fleischmann, N. Schaper, B. Szczyrba, & J. Wildt (Hrsg.), *Neues Handbuch Hochschullehre* (S. 99–120). Berlin: DUZ Verlags- und Medienhaus GmbH.

Microsoft. (2017). https://www.visualstudio.com/de/downloads/

Moody, G. B. (2018). *WFDB applications guide* (10. Aufl.). Harvard-MIT Division of Health Sciences and Technology. http://physionet.org/physiotools/wag/wag.pdf

Object Management Group. (2018). *Unified modeling language (TM)* (version 2.5.1. Aufl.). Object Management Group. http://www.uml.org/

Pan, J., & Tompkins, W. J. (1985). A real-time QRS detection algorithm. *IEEE Transactions Biomedical Engineering, BME-32,* 230–236.

Papula, L. (2014). *Mathematik für Ingenieure und Naturwissenschaftler: Bd. 1. Ein Lehr- und Arbeitsbuch für das Grundstudium.* Springer Vieweg.

Papula, L. (2015). *Mathematik für Ingenieure und Naturwissenschaftler: Bd. 2. Ein Lehr- und Arbeitsbuch für das Grundstudium.* Springer Vieweg.

Rosettacode. (2019). https://rosettacode.org/wiki/Fast_Fourier_transform

Smith, S. W. (1997). *The scientist and engineer's guide to digital signal processing.* California Technical Publishing.

Thakor, N. V., Webster, J. G., & Tompkins, W. J. (1983). Optimal QRS detector. *Medical and Biological Engineering Computing, 21,* 343–350.

Printed in the United States
by Baker & Taylor Publisher Services